Endangered Medicinal Plants

Other published books of the author A.B. Chaudhuri

- Sundarban's Mangrove - Wildlife and Ecology (1985)
- Environmental & Resources of Tropical & Temperate Forests (1990).
- Environmental and Herb Shrub Flora (1991)
- Himalayan Ecology and Environment (1992)
- Mine Environment and Management (1992)
- Plants, Animals and Man (1992)
- Tree and the Environment (1993)
- Forest Plants of Eastern India (1993)
- Sundarbans Mangrove – India (1995)
- Wetland Ecology (1996)
- Biodiversity Endangered (2002)
- Megadiversity Conservation (2003)
- Project Tiger Reserves (2000)
- Plants of Jalpaiguri and Darjeeling (2000)
- Wildlife and Ground Flora (2000)
- Golden Triangle of Eco-education Tourism (2004)
- Fish and the Scientist (2000)
- Mach Niye (in Bengali) (2000)
- Flora & Fauna of Andaman & Nicobar Islands (in Bengali) (2003)
- Deforestation & Perilous Land Degradation (Foresters Unique Role in Amelioration) (2005)
- Biodiversity of Sundarbans and Andaman Mangroves (under publication)
- Wetlands - A Vanishing Resource in India (under publication)
- Forests, Environment and Man (under publication)

Endangered Medicinal Plants

by
A.B. Chaudhuri
Ex-Director, Forest Survey of India

2014
DAYA PUBLISHING HOUSE®
A Division of
Astral International Pvt. Ltd.
New Delhi - 110 002

First Published, 2007

Reprinted, 2014

ISBN 978-81-7035-441-3 (Hardbound)

Publisher's note:

Every possible effort has been made to ensure that the information contained in this book is accurate at the time of going to press, and the publisher and author cannot accept responsibility for any errors or omissions, however caused. No responsibility for loss or damage occasioned to any person acting, or refraining from action, as a result of the material in this publication can be accepted by the editor, the publisher or the author. The Publisher is not associated with any product or vendor mentioned in the book. The contents of this work are intended to further general scientific research, understanding and discussion only. Readers should consult with a specialist where appropriate.

Every effort has been made to trace the owners of copyright material used in this book, if any. The author and the publisher will be grateful for any omission brought to their notice for acknowledgement in the future editions of the book.

Published by : **Daya Publishing House®**
A Division of
Astral International Pvt. Ltd.
– ISO 9001:2008 Certified Company –
81, Darya Ganj, Near Hindi Park,
Delhi Medical Association Road,
New Delhi - 110 002
Phone: 011-4354 9197, 2327 8134
Fax: +91-11-2324 3060
E-mail: info@astralint.com
Website: www.astralint.com

Laser Typesetting : **Classic Computer Services**

Digitally Printed at : **Chawala Offset Printers**

PRINTED IN INDIA

Dedication

Dedicated to Dr. Rakesh Chopra, M.D. and Associate Doctor, Oncologists of Apollo Indraprastha Hospitals of New Delhi whom I have observed deeply involved in the treatment of cancer affected patients very sincerely and ceaselessly by application of latest multi-quality drugs. India needs to conserve its medicinal plants for indepth research on such plants to discover life-saving drugs.

Acknowledgements

The author duly acknowledges the magnanimity of various editors and authors in permitting this writer to use excerpts of their published materials with due acknowledgements.

Thanks are also due to the librarians of Institute of Chemical Biology and State Forest Department for helping this writer with various documents needed for reference work using their libraries.

Preface

I am a forester by profession and a botanist by University degree. Right from my college days I was interested in identification of plant species. During my 34 years of forestry service and thereafter I have all along kept interest in the identification of species. My postings in the State as Working Plans division, Silvicultural division and in the Centre as Chief Co-ordinator Pre-investment Survey of Forest Resources and Director of Forest Survey helped me in the job of knowing species of all categories of plants. I have made a lot of status survey of tree, shrub, herb, flora also.

With these background, I have published a good number of books indicating status of various species. This has given me confidence to state that this country's medicinal plant is in jeopardy and are facing crisis due to various biotic and abiotic causes.

India is a vast country and the number of plant species is high and therefore my observation may be taken as broad based. This leaves much work to be done by foresters and botanists in future. I have not changed/corrected the nomenclature used by various scientists. Also analysis of Western, Central and Eastern Indian flora has been made in this treatise.

Hope the appropriate authorities will take steps to save India's medicinal plants from utter ruination.

131, N.S.C. Bose Road,
Block-10, Flat - 4,
Kolkata-700 040

A.B. Chaudhuri
Author

Contents

Chapter One

Depletion of Land, Deforestation and Medicinal Plant Crisis

The black and white map respresents a rough status of forest in India about 70 years back. Since 1980 Forest Survey of India and NRSA prepared Forest maps of India with the help of Satellite imagery, but in small scale, it was no possible, therefore,

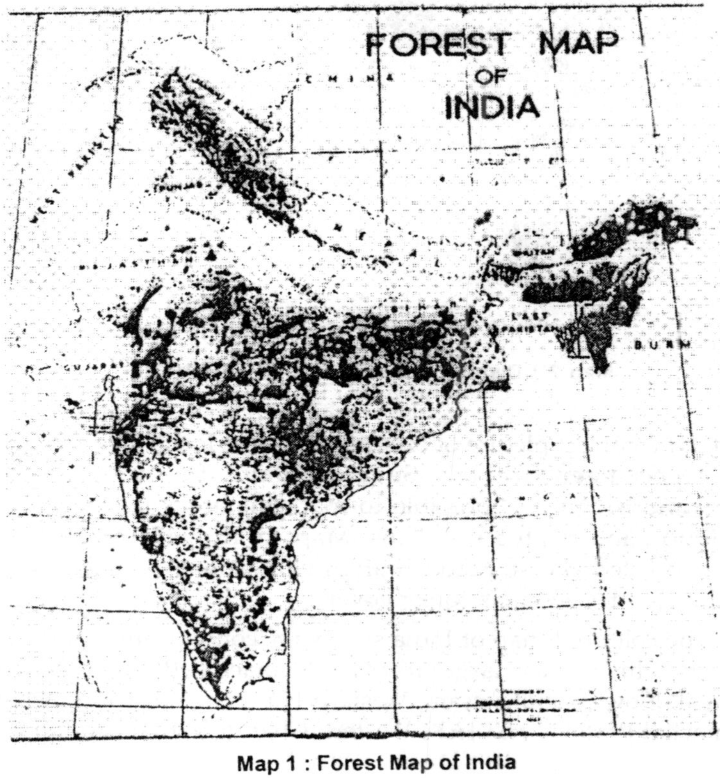

Map 1 : Forest Map of India

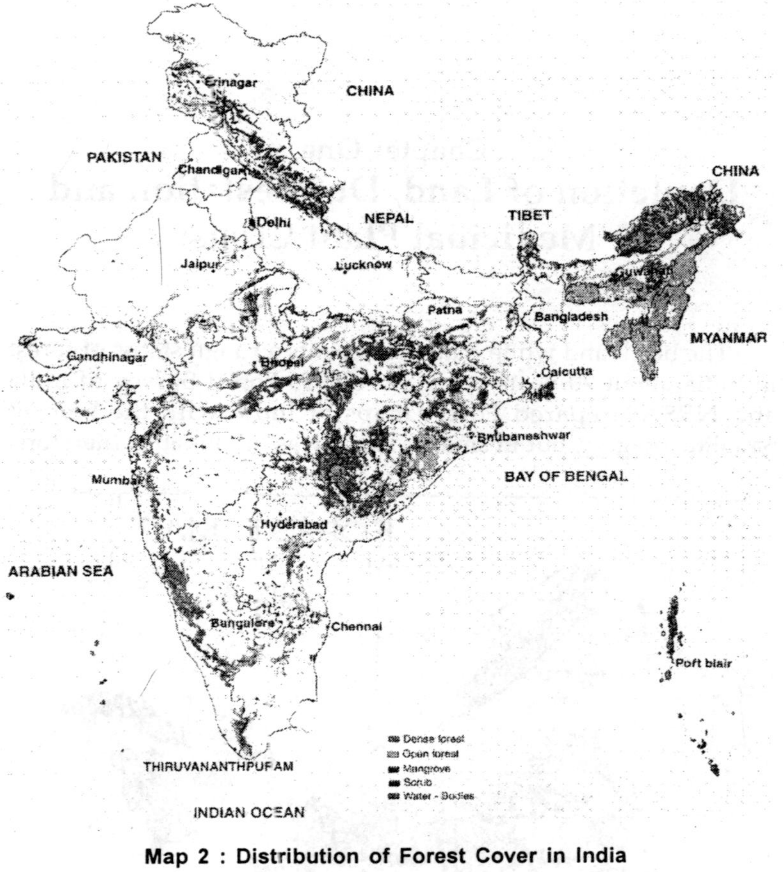

Map 2 : Distribution of Forest Cover in India

to draw a correct picture of existing forests in India. The problem has now been solved. Several maps follows and by superimposition it is possible to mark the areas vulnerable to pollution and related problems. Maps of drought prone area, flood affected and soil eroded areas will expose other problems of the country *vis-à-vis* forest cover.

The coloured map of India shows the concentration of dense forests which cover large areas in Arunachal Pradesh, North Bengal, Nowgaon of Assam, Western Himalayas in Uttaranchal, Himachal Pradesh and J&K, Western Ghat in Karnataka, Kerala,

Bastar in Chattishgarh, Chandrapur in Maharashtra. It also shows the open forests in yellow colour and Scrub in red.

The appended tables (1 and 2) show the forest cover of the country prepared by Forest Survey of India.

Various threats that the country faces has been shown in several charts.

Change in Forest Cover

The change in forest cover of the country, as per the present and the preceding assessments, is given in Table 2a. The net increase in the forest cover of the counry is 3.896 sq.km. The dense forest has increased by 10,098 sq.km. and mangrove by 44 sq.km., whereas open forest has decreased by 6,246 sq.km.

The forest cover in different states and union territories in the present and preceding assessments, along with the data period of the respective assessments, is given in Table 2b. The table reveals that the states of Andhra Pradesh, Arunachal Pradesh, Delhi, Gujarat, Haryana, Himachal Pradesh, Jammu and Kashmir, Karnataka, Madhya Pradesh, Maharashtra, Orissa, Punjab, Rajasthan, Tamil Nadu, Tripura, Uttar Pradesh and West Bengal have registered increase in the forest cover. The States and the Union Territories showing decrease in forest cover are Assam, Bihar, Goa, Kerala, Manipur, Meghalaya, Mizoram, Nagaland, Sikkim, Andaman & Nicobar Islands and Dadra & Nagar Haveli. In Chandigarh and Daman & Diu there was no change in the forest cover during these two assessments.

It is to be noted that the period during which the change in the forest cover has occurred is not uniform in all the states but it varies between 2 to 5 years, because satellite data used in the interpretation are of different dates. The average difference at the national level is about 3 years. In Andhra Pradesh, net increase of 939 sq.km. in forest cover has occurred in 5 years (1993-98), whereas in Mizoram, loss of 43 sq.km. has occurred in 4 years (1994-98). It needs to be underlined that in case of 11 states where the method of interpretation has changed from visual to digital, the present assessment, as such, is not comparable to the preceding assessment because of change in the scale of interpretation. The area of forest cover in such cases has been suitably transformed from 1:50,000 to 1:250,000 scale to

Table 1.1 : Extent of Dense forest, Open forest and Mangrove in States/UTs

State/UT	Dense forest	Open forest	Mangrove	Total forest cover	Per cent of geographic area	Scrub
Andhra Pradesh	24,190	19,642	397	44.229	16.08	9.559
Arunachal Pradesh	57,756	11,091	0	68.847	82.21	104
Assam	14,517	9,171	0	23,688	30.20	324
Bihar	13,274	13,200	0	26,474	15.23	1,914
Delhi	35	53	0	88	5.93	3
Goa	995	251	5	1,251	33.79	16
Gujarat	6,430	5,504	1,031	12,965	6.61	2,948
Haryana	449	515	0	964	2.18	191
Himachal Pradesh	9,120	3,962	0	13,082	23.50	566
Jammu & Kashmir	11,019	9,422	3	20,441	9.20	3,089
Karnataka	24,832	7,632	0	32,467	16.93	4,489
Kerala	8,429	1,894	0	10,323	26.56	91
Madhya Pradesh	81,619	50,211	0	131,830	29.73	3,853
Maharashtra	26,613	19,951	108	46,672	15.17	7,160
Manipur	5,936	11,448	0	17,384	77.86	177
Meghalaya	5,925	9,708	0	15,633	69.70	261
Mizoram	3,786	14,552	0	18,338	86.99	125
Nagaland	5,137	9,027	0	14,164	85.43	14

contd...

Table 1.1 – *contd...*

State/UT	Dense forest	Open forest	Mangrove	Total forest cover	Per cent of geographic area	Scrub
Orissa	26,073	20,745	215	47,033	30.21	5,439
Punjab	517	895	0	1,412	2.80	107
Rajasthan	4,309	9,562	0	13,871	4.05	6,921
Sikkim	2,363	755	0	3,118	43.94	386
Tamil Nadu	8,659	8,398	21	17,078	13.13	2,836
Tripura	2,228	3,517	0	5,745	54.79	38
Uttar Pradesh	22,902	11,114	0	34,016	11.55	1,177
West Bengal	3,565	2,672	2,125	8,362	9.42	98
Andaman & Nicobar Islands	6,515	125	966	7,606	92.21	0
Chandigarh	6	1	0	7	6.14	0
Dadra & Nagar Haveli	159	43	0	202	41.14	10
Daman & Diu	0	3	0	3	2.68	0
Lakshdweep	0	0	0	0	0	0
Pondicherry	0	0	0	0	0	0
Total	**377,358**	**255,064**	**4,871**	**637,293**	**19.39**	**51,896**

Table 1.2a : Class-wise change in forest cover

(sq.km.)

Class	Assessment 1999	Assessment 1997	Change
Dence forest	377,358	367,260	+10.098
Open forest	255,064	261,310	-6.246
Mangrove	4,871	4,827	+44
Total	**637,293**	**633,397**	**+3.896**

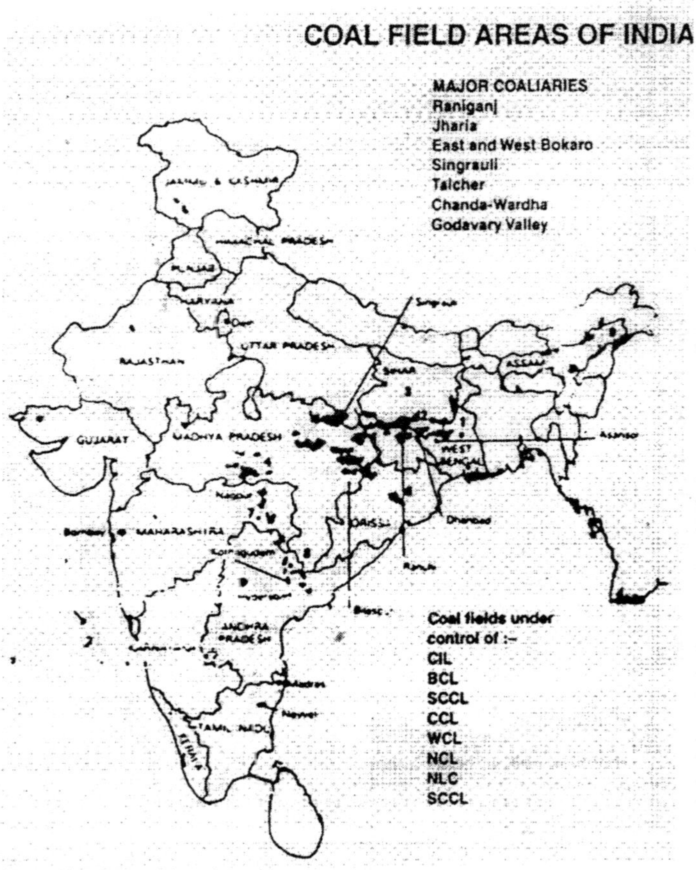

Map 3 : Coal field areas of India

Table 1.2b : Comparative account of forest cover of States/UTs in 1997 and 1999 assessments

State/UT	1997 Assessment		1999 Assessment		Change
	Data period	Forest cover (sq.km.)	Data period	Forest cover (sq.km.)	(sq.km.)
Andhra Pradesh	Oct. 93	43,290	Nov. 98-Jan. 99	44,229	+939
Arunachal Pradesh	Nov. 94 & Nov. 95	68,602	Dec. 98-Feb. 99	68,847	+245
Assam	Dec. 93, Apr. & Nov. 94	23,824	Dec. 98	23,688	-136
Bihar	Nov.-Dec. 94	26,524	Oct. 96-Jan. 97	26,474	-50
Delhi	Oct.-Nov. 92 & 94	26	Oct.-Nov. 98	88	+62
Goa	Dec. 93 & 94	1,252	Dec. 95-Jan. 96	1,251	-1
Gujarat	Oct.-Dec. 94	12,578	Oct.-Dec. 96 & 97	12,965	+387
Haryana	Oct. 92 & Oct.-Nov. 94	604	Nov.-Dec. 96	964	+360
Himachal Pradesh	Oct. 94 & Nov. 95	12,521	Oct.-Dec. 98	13,082	+561
Jammu & Kashmir	Nov. 94 & 95	20,440	Nov.-Dec. 96 & Sep.-Oct. 97	20.441	-1
Karnataka	Dec. 93-Jan. 94	32,403	Dec. 95-Jan. 96	32,467	+64
Kerala	Dec. 95	10,334	Jan. & Mar. 96	10,323	-11
Madhya Pradesh	Oct.-Nov. 94	131,195	Oct.-Dec. 96	131,830	+635
Maharashtra	Oct.-Nov. 93	46,143	Oct.-Nov. 96	46,672	+529
Manipur	Dec. 93-Feb. 94	17,418	Dec. 98	17,384	-34
Meghalaya	Dec. 93	15,657	Dec. 98	15,633	-24
Mizoram	Dec. 93-Feb. 94	18,775	Dec. 98	18,338	-437

contd...

Table 1.2b – *contd...*

State/UT	1997 Assessment		1999 Assessment		Change
	Data period	Forest cover (sq.km.)	Data period	Forest cover (sq.km.)	(sq.km.)
Nagaland	Dec. 93-Oct. 94	14,221	Dec. 98	14,164	-57
Orissa	Oct.-Nov 93	46,941	Nov.-Dec. 95	47,033	+92
Punjab	Oct. 94-Jan. 95	1,387	Nov.-Dec. 96	1,412	+25
Rajasthan	Nov.-Dec. 94	13,353	Oct.-Dec. 96	13,871	+518
Sikkim	Sep. 94	3,129	Nov. 98	3,118	-11
Tamil Nadu	Mar. 94	17,064	Jan., Mar. & Sep. 96	17,078	+14
Tripura	Dec. 93	5,546	Dec. 98	5,745	+199
Uttar Pradesh	Oct.-Dec. 94	33,994	Oct.-Dec. 96	34,016	+22
West Bengal	Oct.-Dec. 93	8,349	Dec. 95-Feb. 96 & Dec. 96	8,362	+13
Andaman & Nicobar Islands	Dec. 94	7,613	Mar. 97 & Jan.-Mar. 98	7,606	-7
Chandigarh	Oct.-Nov. 94	7	Jan. 99	7	0
Dadra & Nagar Haveli	Oct.-Dec. 94	204	Nov. 96	202	-2
Daman & Diu	Oct.-Dec. 94	3	Nov. 96 & Oct. 97	3	0
Lakshdweep	–	0	–	0	0
Pondicherry	–	0	–	0	0
Total	**–**	**633,397**	**–**	**637, 293**	**+3,896**

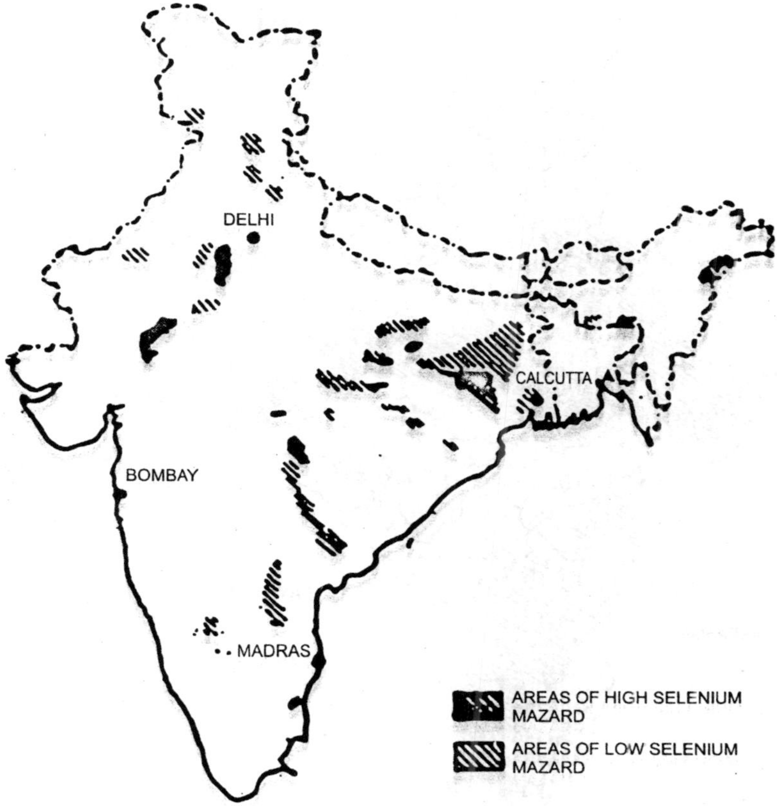

Map 4 : Map showing selenium polluted areas

maintain consistency in comparison. For doing so, digital interpretation was done on both the scales using the same data of the sampled area. A ratio factor was then estimated for each sheet for converting the area of forest cover from 1:50,000 to 1:250,000 scale.

Map 3 showing major coal belts of India. Foresters have a very wide field of works to perform in the form of soil conservation and afforestation to ameliorate the disturbed environment.

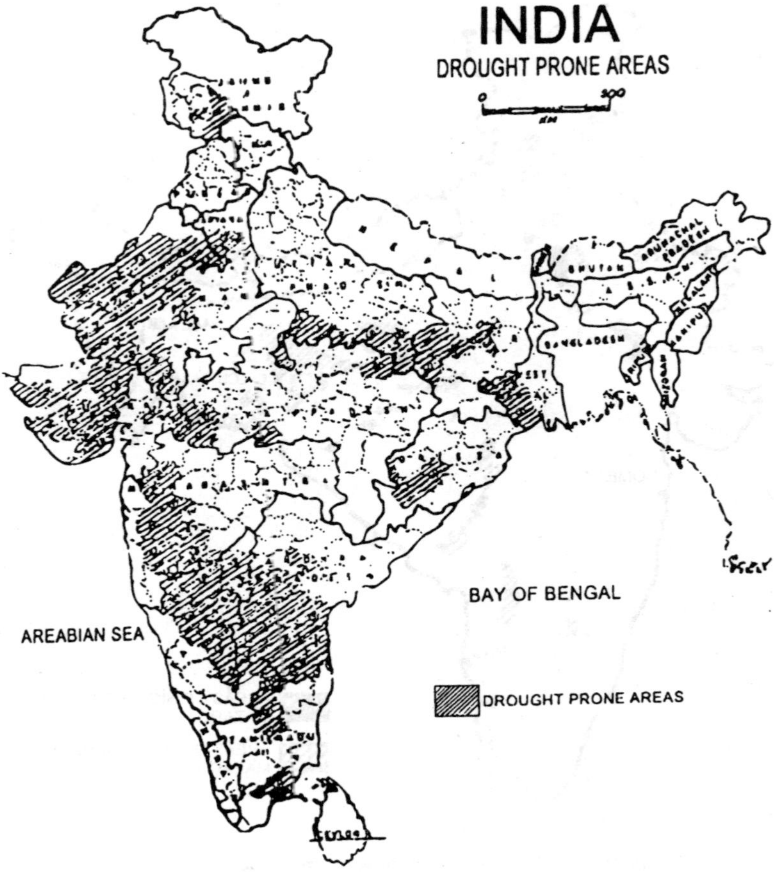

Map 5 : Map of India showing the drought prone areas.

Map 4 may be super imposed on other maps on mine, mineral, thermal power site maps and the depth of the environmental problems assessed.

Map 5 represents the drought prone areas of India which very frequently become the victim of drought and sometimes severe drought conditions.

A rough presentation of Industrial and mining areas of the country has been made on this map. The cumulative affect of all the issues presented in the maps 4-9 has to be assessed and

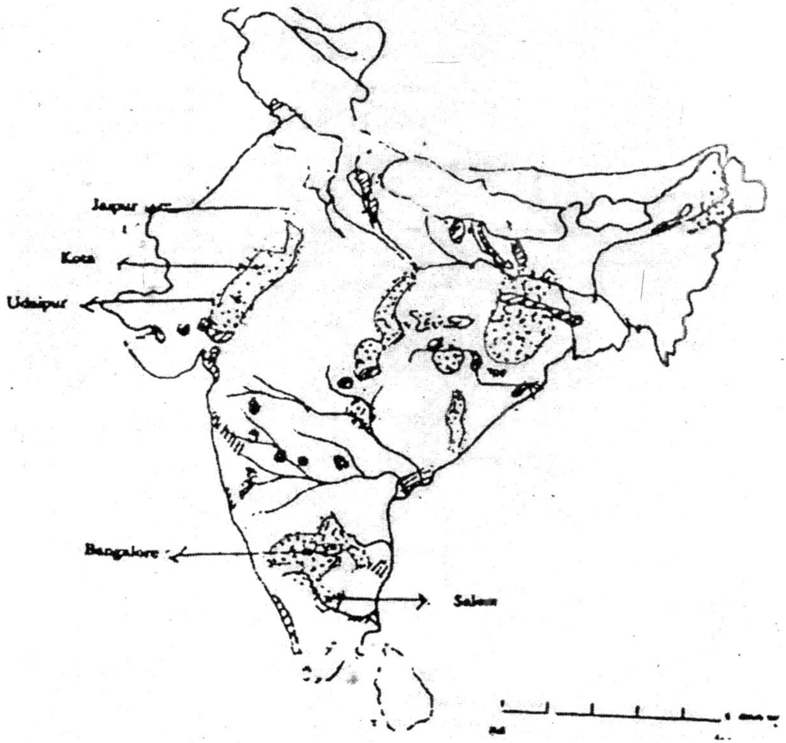

Map 6 : Industrial and mining areas.

remedial methods adopted. Foresters have a great role to play in amelioration.

Map 7 broadly shows the mine areas of this country. The map has to be read along with other maps and steps taken to rehabilitate and ameliorate the environment through various forestry methods.

Of the aforesaid geographical distribution of medicinal plants of India, the richest regions are the Himalayan forests and Western ghat forests. Other areas are either arid, desert or heavily populated regions of the plains.

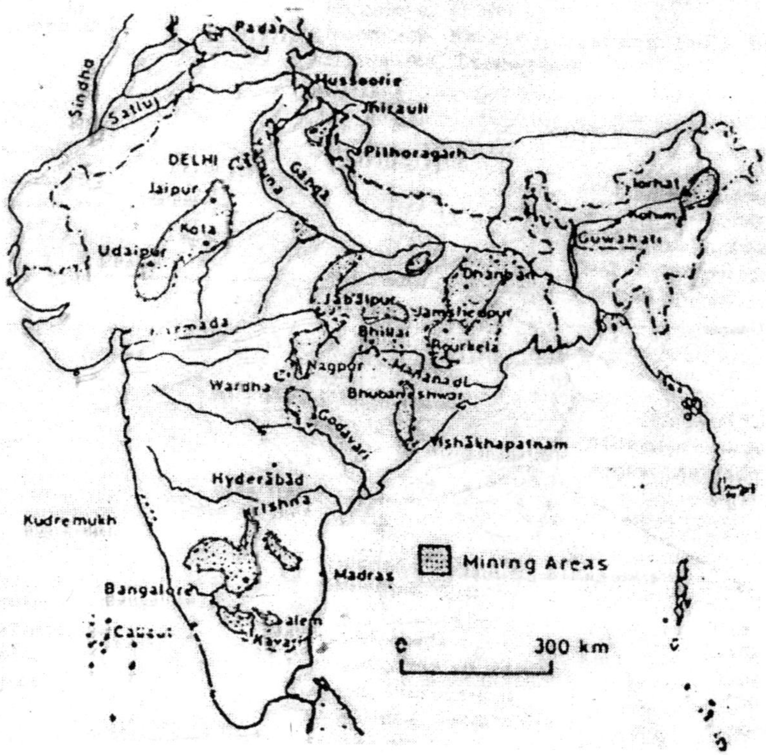

Map 7 : Mining areas of India

India Facing a Perilous Landuse and Environmental Crisis

The plannings of India were done without considering the terrible landuse situation on which the country's future stands. It is like building castles over a soft, cracked and dilapidated foundation that might breakaway anytime like quick sand. The authors drew parallel between the Roman emperor 'Nero' (37–60 AD) (dream of building new city of Rome while he was

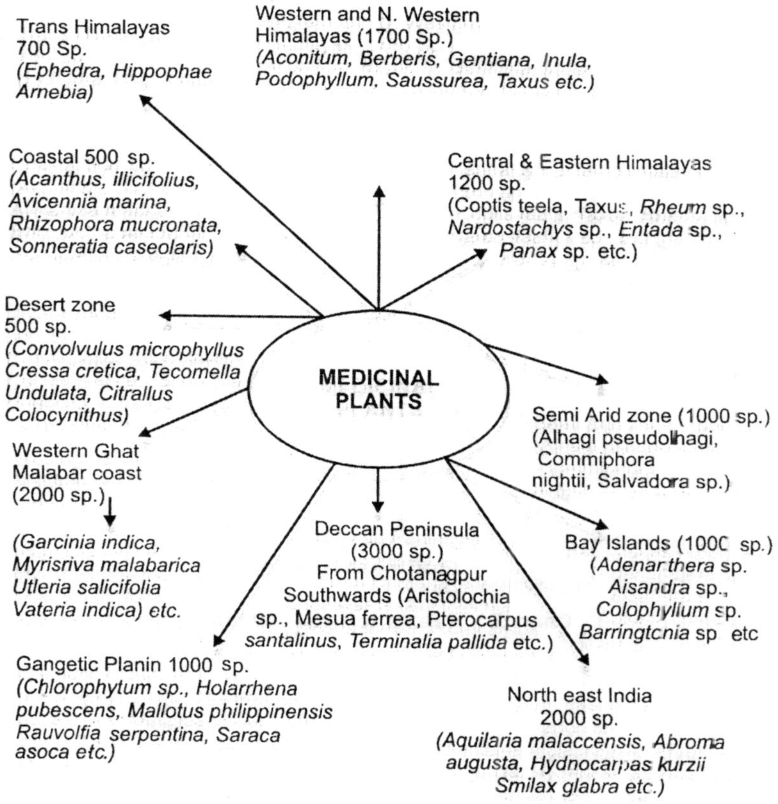

Chart 1 : Biogeographical distribution of medicinal plants.

playing lyre in full view of burning city) and the planning for development of modern India, the latter having been fully oblivious of landuse situation.

Besides these natural calamities India has to tackle the problems of a huge poverty-stricken, illiterate, unemployed mass where even amenities like safe drinking water, education and medical help etc. are lacking.

The author while analysing various facts and figures, documented by the Planning Commission of India, the National Commission of Agriculture and the National Commission on Flood found that they have registered 70% of the total 3.2 million

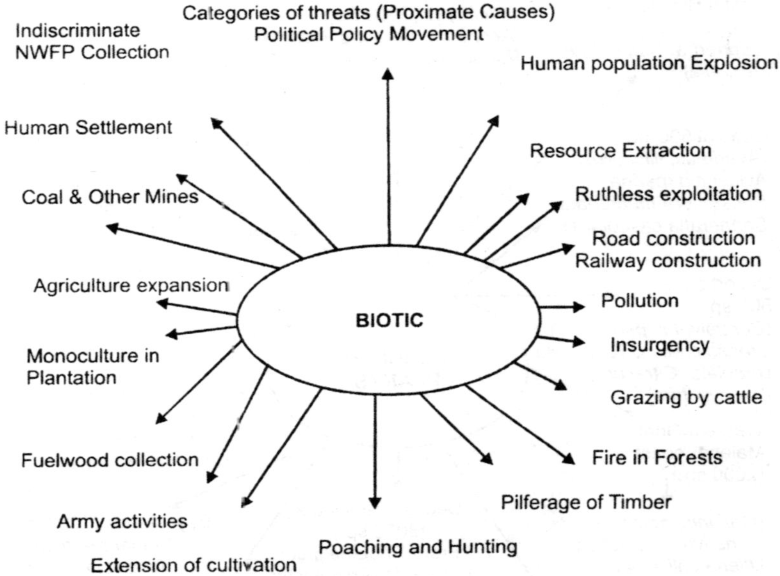

Chart 2 : Threats to India's Nature

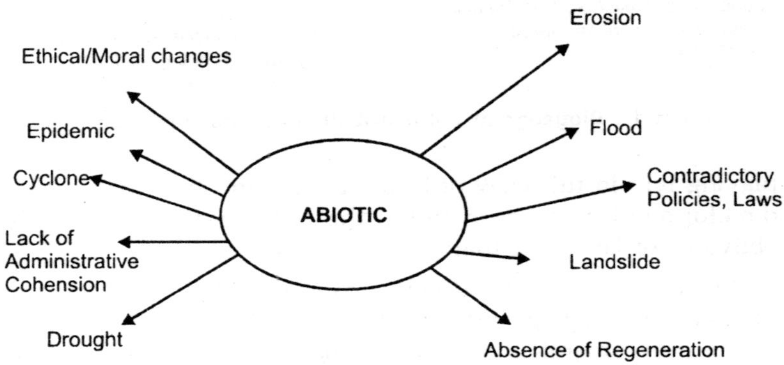

Chart 3 : Root Causes

Chart 4 : Selected study Area
(Comments based on published reports supplemented
from author's field work)

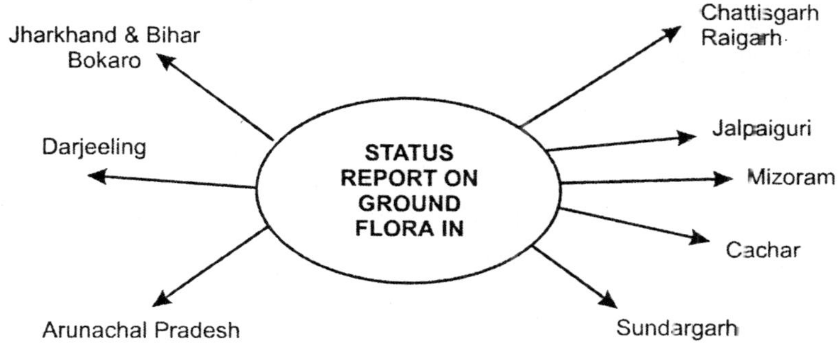

Chart 5

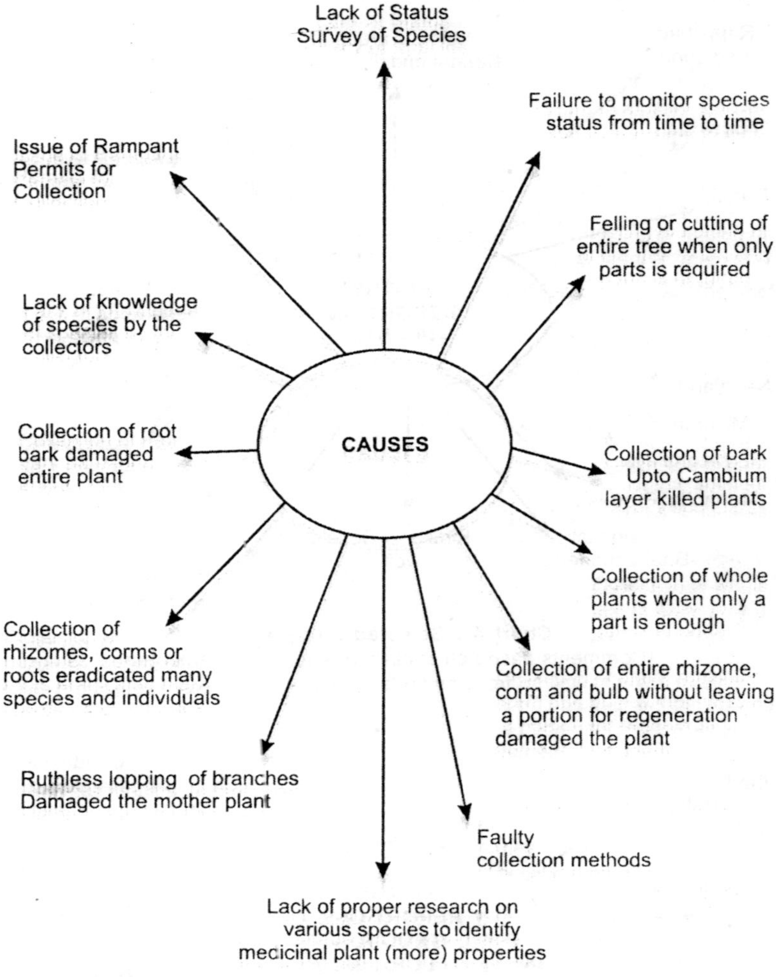

Chart 6 : Other causes of depletion of medicinal plants

sq.km. of land has been reeling under flood, drought and erosion etc. Such landuse condition has to be tackled effectively before India takes up a massive, gigantic and long-duration projects. They have stressed that India needs to create an effective cover of plants by massive afforestation mingled with soil conservation engineering works, reclamation of waterbodies– big or small. Here comes the role of foresters who are well- equipped to create a better India. They have training, expertise and the character built over an experience of about 200 years to rehabilitate depleted soil.

This chapter presents some data on flood, drought, soil, erosion, forest cover, etc. and relevant issues.

Medicinal Plants (A dwindling resource)

The author sincerely feels that the vast reservoir of medicinal plant resources of India is in peril. The readers may peruse the chapters on 'medicinal plants' detailed in the books (i) Biodiversity Endangered (India's Threatened Wildlife and Medicinal Plants) and (ii) Megadiversity Conservation (Flora, Fauna and Medicinal Plants of India's Hot Spots) of the author.

The author has analysed various issues on the subject and has given a broad view of the status of important plants.

The author has stressed the point about qualitative study by the Botanical Survey of India on the flora of various region. Without quantitative studies the status of any plant cannot be ascertained. As such, analysis of ground flora (herbs, shrubs, climbers) which form the bulk of medicinal plants is essential to determine the status of a plants.

Forest Survey of India, has initiated the study of ground flora all over the country which should focus some important features of ground flora.

A Resource that is Fragmenting Rapidly

Indians are fortunate to be blessed with varieties of climate, from tropical to alpine and desert to humid. It has therefore, a large array of vegetation and having more than 20,000 plant species. Most part of this sub-continent falls in the biotic region of tropical deciduous forests and tropical scrub forests. Tropical rain forests occupy a narrow belt in the west coast, north-east

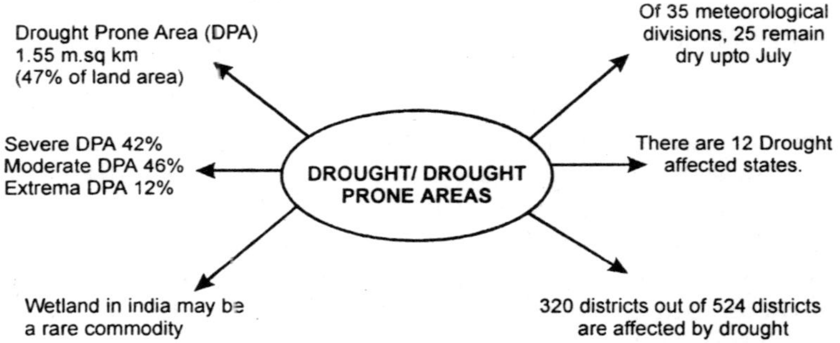

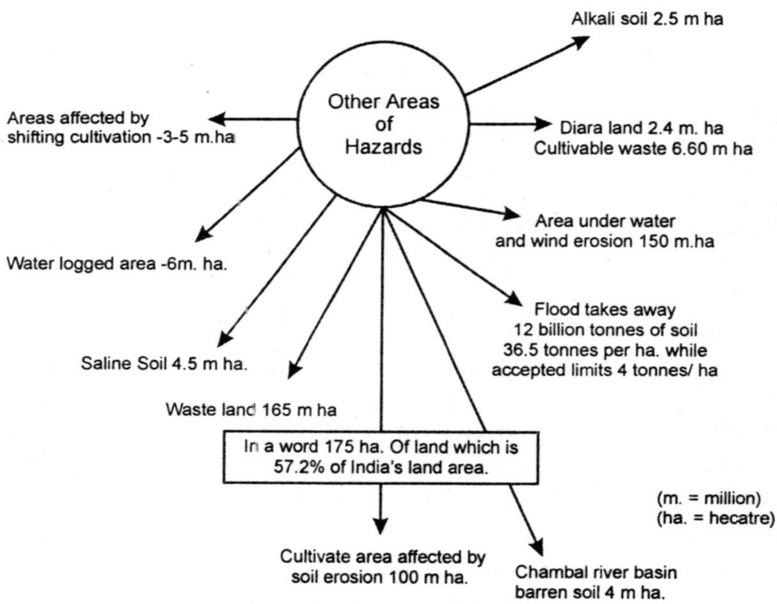

Chart 7 : Perilous land use (Vohra's Assessment)

India and in Andaman and Nicobar Islands. The desert and arid zone vegetation is found in Rajasthan and adjoining areas. Sub-tropical, temperate and alpine forms of vegetation are found in the Himalayas, Nilgiris and other hill regions.

PERILOUS LAND USE (*Contd...*)

Accelerated erosion:

Water erosion	– 90 million ha.
Wind erosion	– 50 million ha.
Seasonal flood	– 12 million ha.
Inundated annually	– 0.7 million ha.
Total land area	– 328 million ha.

Various Hazards : Drought prone area 155 m. sq.km.

- Ravine affected area - 73–69 lakh ha.
- Degraded areas 35 million ha. (Vohra (1978)

FAO 1981 Degraded Forest 15.095 million ha.

- Scrub (no tree)
- – 5.378 m. ha.
- Open forest (scattered trees)
- – 5.393 m. ha.
- Forest, subjected to heavy
- – 4.324 m. ha.
  biotic pressure
- Rate of degraded in 0.8 million ha. per year Khan 1987

Various Hazards :

- Exhausted by over grazing.
- Scarcity of fuel wood and fodder.
- Degradation of agricultural land.
- Damage of water bodies.
- Deprived of humus due to constant burning.
- Repeated flood and soil erosion causes of degradation.
- Increase in population and clearance for agriculture.
- Increased live stock population.
- ..... for fuelwood timber, shifting cultivation.

Indians Horrendous Land Degradation situation

Various arid zones :

Desert zone	–	929000 ha
Arid zone	–	13455200 ha
Semi-arid zone	–	6686000 ha
Sub-humid zone	–	110000 ha
Humid zone	–	211000 ha

Desertification

Arid areas 3200000 ha

(12% of country's area)

The extent of forest and the condition may be perused from maps and tables.

India has 16 major vegetation types which are • Tropical Wet evergreen forests. Tropical Semi-evergreen forests • Tropical moist deciduous forests • Littoral and Swamp forests • Tropical Dry deciduous forests • Tropical thorn forests • Tropical dry-evergreen forests • Sub tropical broad leaved hill forests • Sub-tropical dry evergreen forests • Sub-tropical pine forests • Montane wet temperate forests • Himalayan moist temperate forests • Himalayan dry temperate forests. Sub alpine forests. Moist alpine Scrub forests • Dry alpine Scrub forests.

The Himalayan region is fragile; also the Thar desert, the western and eastern ghats. The Himalayan region covers 50.0 million ha. The ecological stability of the region is not only important for the local population it is equally important for the entire Indo-Gangetic basin. The arid region occupy 31 million ha., 60 per cent of which lie in Rajasthan. Grazing, fire, uncontrolled collection of wood continue round the year.

The extent of erosion may be visualised from the media report (Times of India dt. 11.11.05) quoted in this chapter.

Crisis Facing Medicinal Plants

The present conception of people over a wide circle is that India's floral and faunal resources are considered very rich. In the past sixty years the biotal scenario has undergone a drastic change due to uncontrolled biotic factor; some abiotic factors have also aggravated the situation. The Department of Forests was created since the sixties of nineteenth century and on the recommendation of the Government of India enacted Wildlife Protection Acts for conservation of 'Birds', 'Elephants' and 'Rhinoceros' and lately Wild Life (Protection) Act of 1972 for protection of all biotal species those are threatened of existence. This Act of 1972 takes care to protect various categories of flora and fauna in consideration to their present status. In various management plans drawn for Forest Divisions adequate care had been taken to preserve threatened tree species; but little care was taken for the preservation of herb and shrub flora which form bulk medicinal plant materials.

India has more than 1500 tree species of which only few are regenerating. Over-exploitation, selective exploitation, grazing and fire are threatening their survival. Trees which occupy the topmost layer of a forest may not serve a foolproof function against various atmospheric hazards; top middle and bottom, the middle layer has the maximum number of species; next is the bottom layer and it is hardly much forest left in the country where the bottom and middle storey trees stand. Even the top layers have been stripped off of many tree species. All these give a dismal picture of the state of vegetation in this country.

Of all life-forms of plants in India, the number of herb species out number others. Roughly, the number of shrub species is about double the number of tree species while the number of herbs is about four times the Shrubs.

Over-exploitation of tree crops in all the three storeys cause considerable physical damage to shrubs; these shrubs are essential for soil binding and create a micro climate for the survival of young plant regeneration; besides, they help litter formation and conservation of water.

Shrubs and young regeneration of other plant species suffer irreparable losses due to repeated grazing and fire. These latter two factors are directly responsible for eradication of plant species where only those with underground stem survive. The authors' research reveal that only about 40 to 50 per cent of the species survive such depredation; also the density of occurrence of species is affected. Vast tracts of forests where grazing and fire have a free play have now very coarse grass and a few fire hardy shrub and some annual herb species.

About ninety species of shrubs are conspicuous in the temperate hills of India which have colourful. flowers; they also grow in profusion. Special mention may be made of rhododendron and roses. In the plains, however, more than one hundred species occur but they do not have conspicuous flowers. In eastern India there are at least a hundred species of fodder shrub; it is about two thirds in other parts of the country. There are at least forty genera of shrubs yielding edible fruits which attract avifauna, animals .and men alike.

A survey made by the authors all over the country has revealed that only fifty species of shrubs are commonly found. Lantana, Eupatorium, Clerodendrum, Calotropis, Cassia, Carissa, Ipomoea, Capparis, etc. are found conspicuous and are occurring in profusion all over the country. It should be considered an ominous indication of systematic disappearance of species primarily due to anthropogenic factors. The number of shrub species in the country is about 4500 (tree species about 1500); but wide occurrence of only 50 species is hardly to be believed which is too meager.

Vast shrub resources of this country have not been studied from medicinal, aesthetic and environmental point of view. This country has shrub flora occurring on wide ranges of alkaline, saline, estuarine, sandy soils in varied climatic conditions. The time is ripe to select typical shrubs that have conspicuous flowers, wide range of foliage patterns and shapes and plant them in arboreta, parks and gardens. Some shrubs are indispensable ingredients of parks, gardens, residential quarters premises and in environmental planting to arrest air pollution.

An attempt has been made to list some common and conspicuous shrubs occurring all over the country (state-wise) and to mark some common medicinal plants. Although some of these shrubs are obnoxious weeds (Eupatorium, Cleistantus, Lantana, Calotropis, Euphorbia, Jatropha etc.), quite a good number occur on dry rocky and sandy areas (Euphorbia, Opuntia, Agave, etc.) which provide shade, protect soil, shelter birds and bear fruits for man and animal besides, providing plants of medicinal values. Vast avifaunal species in the country side derive a lot of food and seek shelter in such shrub flora.

A quotation from Times of India, Nov. 11, 2005 by Chandrica Mago reads:

Alarm bells are ringing for "sons of the soil". The government says the country is losing its soil at an unacceptable rate. The country's average soil erosion rate is 16 tonnes per hectare per annum – more than three times the acceptable norm. The figure goes as high as 80 tonnes in the Himalayas. Scientists say the figure for Hoshiarpur is the highest in the country, at a whopping 400 tonnes a hectare every year. It takes a million years to replace topsoil.

In areas losing soil, the impacts can be drastic. Productivity comes down. Agriculture minister Sharad Pawar recently estimated that soil erosion leads to a loss of about 8–9 million tones of valuable nutrients. Second, the siltation rate goes up in crucial water reservoirs. It is reducing storage capacity by 2% each year. This impacts water availability on all fronts. Third, certain grass and tree species are vanishing. Down stream, silt in rivers is forcing them to change course.

Each earthquake creates a more suitable ground for erosion. There are about 300 small tremors in a year across the mountains – we may not feel them but the soil does. Melting glaciers are leading to more erosion in specific locations, says J.S. Samra, Indian Council of Agricultural Research deputy director-general.

The problem is getting worse since there are more people and livestock crammed into the same amount of land. Trees are being cut, grasses are vanishing, leaving bald patches vulnerable to erosion when the rain comes down hard. It doesn't matter how much it rains in a year – what matters is the intensity of rain, the sheer force which loosens the soil. One rain drop falls at 16 metres a second, says Samra "You can actually see the holes it leaves in theground. Yet, it's a problem little talked of outside scientific circles." Samra says, soil erosion across the country ranges from 2–80 tonnes per hectare every year, with rainfed areas suffering more than irrigated plains. An acceptable range is 5 tonnes., The irrigated Indo-Gangetic plain, where water is trapped, has a rate of less than 2 tonnes. After the Himalayas, the highest rates of 40–50 tonnes show up in the Nilgiris and the western and eastern Ghat.

If there are forests and even better, grass, the ground holds. Cultivated rather than forested slopes are more vulnerable to erosion. There are solutions and programmes to check erosion - but they have made limited impact. The best solution, says Samra, is to treat catchments, creating trenches to hold the water and reduce the speed of flow. If water velocity increases two times, its cutting capacity goes up four times and its carrying capacity, eight times. Too much erosion can muddy the waters so much that fish starve for oxygen. Bihar is a victim of uncontrolled erosion as water floods in from Nepal, Rajasthan, of course has its own set of problems – wind, not water, erosion."

Destructive forces have also affected these forests. Refugee rehabilitation (25,000 East Bengal refugees) has been responsible for bulk destruction and clear felling of about 1,0000 ha. Has augmented substantial depletion of the forests.

Selection felling done to ensure natural regeneration has opened up extensive areas of forests which have lost the character of tropical forests. Many areas have been exposed to erosion. Most significant change and degradation has been caused by severe cyclone which is a regular feature of Andaman forests. Hundreds of square kilometers of forests have been uprooted and damaged (shallow root system and heavy crown of trees). Ground flora has also been disrupted.

All these derogatory factors combine to bring a synergetic destructive presence on the soil of the country.

Chapter Two

Medicinal Plants of India : An Observation

This chapter summarises the findings of a group of experts observation on the subject.

This chapter enumerates broadly the following information:

- Richness of flora.
- Past history.
- Use of herbal drugs in India.
- Plants used in drug industry.
- Essential oils.
- Phyto-pharmaceuticals.
- Export market.
- Commercial value of medicinal plants used as drugs.
- Trade market.
- Threat to plant biodiversity.
- In-situ conservation.
- Ex-situ conservation.
- Cultivation.
- Action Plan.

The author acted as a Chairperson in Forest Bio-Diversity Section of a study initiated by the Department of Environment, Govt. of West Bengal. His findings along with those of four other Chairpersons findings were compiled in a report entitled "Biodiversity Strategy and Action Plan for West Bengal". Some relevant paragraphs from that report are presented taking the liberty that the author was a part of the compilers of the report.

The study of medicinal plants has assumed great importance in India and abroad. In various states of India, no proper inventory of medicinal plant have been prepared. So far the work done in the field is in the form of scattered or isolated efforts, aiming mostly towards medicinal use of the items. In fact, preparation of an inventory alone would not be able to meet the growing need of planned utilization of medicinal plant resources within the states. Some basic information on the medicinal plant species is also required to design and implement proper management plan for the extraction and use of medicinal plants in future. It is also important to assess the level and extent of exploitation of medicinal plant from the wild stock and to put in place necessary checks. The knowledge in this field would enable introduction of timely cultivation in necessary cases and maintain the required balance between proper sustainable use and exploitation of limited natural resources.

Status of Medicinal Plants

India harbours a wide range of medicinal and aromatic plants mostly used in Ayurvedic, Unani, Siddha, Homeopathic, Allopathic and other alternate medicinal practices such as folk medicine, home remedies, household remedies naturopathy, tantra-therapy, Amchi and tribal medicine. The plants used in alternate medicine are awaiting a touch of modern knowledge.

A rough conception is that out of 17,500 flowering plant species known from India, more than 4000 species are used as medicinal plants of which 300 species yield gum and dyes and about 100 species yield essential oils and are used as raw materials in drug industry. About 200 drugs are of animal and mineral origin.

Keeping in mind this perspective it can be mentioned that the Rigveda (4500–1600 B.C.) an old repository of human knowledge, quoted 67 plants of medicinal value. The Atharva Veda and the Yajurveda mention 200 and 81 species of plants of therapeutic value. Charak (1000 B.C.) and Sushruta (800 B.C.) quoted 341 and 395 medicinal plants respectively in their Samhitas. Ethno-botanical studies have added about 1680 species of medicinal value in the existing list of medicinal plants in India.

In last 75 years the basic active principles like salicylic acid. alkaloids, sapouins, glycosides, steroid compounds have been isolated from higher group of plants except Ephidrin from Ephedra (a gymnosperm), Taxol from Taxus (Gymnosperm), new alkaloid from Lycopodium, ergot from Claviceps fungus and other antibiotic from fungi groups. During 1950, only 2000 alkaloid was known to the world. By 1970, the number increased to 4000, and in 2000 A.D. it has been estimated to be about 10,000 alkaloids, of which 5000 are chemically less known.

The discovery of hallucinogenic use of morning glory has stimulated further studies of Datura, Canabis, and Papaver in India. The hallucinogenic drugs are very much in demand in today's world. Very recently steroidal sapogenious substance has also attracted phyto-chemists to give more attention in their research.

India uses herbal drugs utilizing nearly 450 plant species in different formulations (Kapoor and Mitra, 1979).

In this context three orchids can be mentioned which are used in traditional medicine and at present endangered or vulnerable (Lucas and Syng. 1978). Paphiopedium druryi is native to Kerala but is endangered or extinct from the wild due to forest fires and excessive collection. There is an indication of its use in Ayurvedic medicine and that it might contain useful alkaloids. Various authorities have indicated its decline in the wild state (U.C. Pradhan, 1975, 1976, 1977; Mammen and Mammen, 1974). Dendrobium pauciflorum is endangered, possibly extinct from West Bengal and Sikkim in areas open to tree felling. It can be rediscovered in the wild, it is likely to contain some alkaloid of potential value. Diplomaris hirsute, which possibly contains useful alkaloids in the tubers, is restricted to very few numbers in West Bengal in a region vulnerable to landslides. Its decrease in number has been pointed out by G.M. Pradhan (1976) and Varman and Sahni (1976). Another. species Dendrobium nobile, deserves mention. Occurring in the Himalayas regions of India and China, it is a source of dendrobine, a principal alkaloid and is exported in tones from China in dehydrated form (Pemupahishey, 1974). The abundance of this species in India is decreasing rapidly and is a matter of serious concern (Santapu, 1970, Kataki, 1976).

An examination of the traditional systems of medicine – Ayurveda and Unani systems indicated that Rauvolfia serpentina used by medicine men of India, Sri Lanka, Nepal and Mayanmar for treatment of insanity. This is possibly associated with hypertensive encephalopathy. Most European physicians were very skeptical of the purported properties of the plant. However in 1952, the alkaloid reserpine was isolated, thus confirming the plant's value. Since then the alkaloid extract as well as purified alkaloids of Rauvolfia serpentina have become very important in the treatment to control hypertension.

Medicinal Plants Based Drug Industry in General

Medicinal plants based drug industry has four major segments :

- Plant drugs for WHO recognized Indian systems of medicine.
- Over the country non-prescription items involving plant parts, extracts and galenicals.
- Essential oils.
- Phyto-pharmaceuticals.

Besides these, plant derived materials including fats, oils, waxes, latex, pectins, resins and oleoresins, gums and other exudates, vegetable dyes and tannins, lignin, cellulose, starch, hydrocarbons and many other bio-chemical compounds (Pryde *et al.* 1981; Schultes, 1980) are important primary items. On top of these primary products are numerous secondary compounds and chemical intermediates, including sterols, alcohols, alkaloids, resins and esters.

These diverse categories are almost certain to expand as scientific research comes up with more products from tropical plant – not only items more in number, but materials with greater complexity and novelty. In the light of speedy expansion of the global chemical industry, it is clear that we shall need much more additional supplies of organic chemicals.

The chemical industry continues to drive most of its feedstock from fossil fuel. But due to the oil cartel *i.e.,* the Organisation of Petroleum Exporting Countries (OPEC), many petrochemicals have increased 6–7 times in price whereas vegetable fats and oils have not even doubled in price.

Fortunately, almost all petrochemicals can now be replaced by phyto-chemicals.

Medicinal Plants Based Drug Industries in Indian System of Medicine

According to Handa, in 1992, there are more than 6,780 pharmacies in Indian systems of medicine of which 551 are on loan license and the remaining are from D. In West Bengal,over 642 pharmacies are in existence of which 620 are for the production of Ayurvedic drugs and rest 22 for Unani drugs. All these are licensed pharmacies. Besides these, there are many small manufacturing units using medicinal plants and thousand of vaidyas, and hakims preparing their own drugs from various medicinal plants.

Plant Parts Extracts

The direct utilization of plant material is a feature of traditional systems of medicine not only in India, but also in developed countries like Europe, U.S.A., Western countries now a days are very much fond of herbal formulations of health, food, preparation of decoctions, tinctures, etc. And total extracts of plants also from a part of many pharmacopoeias of the world. The current trend of medicinal plants based drug is to procure standardized extracts of plants as raw material.

Essential Oils from Plants

The essential oil industry was traditionally a cottage industry in India. Since 1947, a number of industrial companies have been established for large-scale production of essential oils, oleoresins and perfumes. The essential oils, oleoresins from plants being produced in India include ajawain oil, celery oil, citronella oil, lavender oil, eucalyptus oil, geranium oil, cedarwood oil, sandalwood oil, lemongrass oil, vetiver grass root oil, mentha oil, turpentine oil and resin from pines. This forms a sizeable and well-established industry in India having annual oil production of about 50,000 tons. [I]-lonone from lemongrass oil for perfumery, and [I]-lonone from vitamin A synthesis are produced in India. Before 1960, Menthol was not produced in India, but the introduction of Japanese mint – Mentha arvensis L. (Lamiaceae) and subsequent improvements there upon have enabled India to produce more than 600 tons of menthol and till date top the world market in export of natural menthol.

Limonene is the bye-product of citrus industry. Though turpentine oil and eucalyptus oil also yield limonene, but the best economically cheap raw materials is the discarded orange and lemon neel which are being used by Brazilan phyto-chemical industries. India, particularly West Bengal has not yet tapped this source for limonene production.

Phyto-pharmaceuticals

Since the independence of India, the production of plant based modern drugs in India and West Bengal in particular was mostly confined to quinine, from cinchona. During the last five decades, bulk production of plant based modern drugs has become an important segment of Indian pharmaceutical industry. Some of the phyto-pharmaceuticals which are produced in India include Morphine, Codeine, Papaverine, Thebaine, Emetine, Quinine, Quinidine, Digoxin, Caffeine, Hyoscine, Hyoscyamine, Xanthotoxin, Psoralen, Colchicine, Rutin, Berberine, Vinblastine, Vincristine, Nicotine, Strychnine, Brucine, Ergot, Alkaloids, Senna, Glycosides, Pyrethroids, Podophyllotoxin resin, Steroid compounds etc. Phyto-pharmaceuticals for which technology has been developed for undertaking large-scale production include L-dopa from Mucuna, pruriens, Ajmaline and Ajmalicine from Rauvolfia serpentina and Catharanthus roseus respectively and B-acetyl glycyrrhetic acid from Glycyrrhiza glabra. Medicinal plant based drug industries have started facing and will face infuture, dwindling supply of plant materials from natural resources. So promoting cultivation of medicinal plants which are being extensively used by the industry, will help to solve the problem of the industry.

Status of India in Herbal Export Market

India's share in world herbal medicine market is mere 11% and the total market from Indian systems of medicine and Homeopathy is about US $ 1 billion – stated by G.C. Burman, Chairman of Dabur India Ltd.

India is losing out on the herbal export front, while China earns US $ 5 billion annually from herbal trade which is five times more than India's export turnover. Intensive efforts are being made in India to boost its export of traditional medicines to match at least half of China's current export level by 2010. Indian exporters feel that given an effective and sustained push, its traditional medicine export would be able to make significant profit, almost equal to that earned from export of computer software.

Traditional medicine manufacture and practices are largely based on plant products. The international market for traditional products is amount US $ 62 billion and is expected to reach US $ 5 trillion by 2050.

Commercial Value of Medicinal Plants and Plant-based Drugs

Medicinal plant diversity has its economic value and it has local, regional, national and international implications like patent right, intellectual property right etc. It has also has an alternative value, an intrinsic value (Pearce and Moran, 1994), which is academic/scientific, that is unrelated to direct human use.

Medicinal plant species are used for therapeutic purpose in three ways:

- As traditional and alternative medicines singly or in formulations, such as those prepared and dispensed by traditional and alternate medicine practitioners, which may or may not attract a market price.
- As commercial products, dispensed by prescription or over the counter sales, such as patented/licensed medical products of Allopathy, Homeopathy, or traditional systems medicine.
- As bazaar medicine singly or in formulations.

All these usually have economic value. For the lack of adequate and appropriate data, it is near impossible to evaluate the returns from the first and the third categories. The economic value of plant based drugs, therefore, largely rests on the second category uses. It is estimated that in rich countries, 25% of all medical drugs are based on plants and their derivatives (Principe, 1991). In the poor world, this is about 75%.

The economic value of a particular medicinal plant depends upon a number of factors as follows:

- Certain plant species are used in a large number of formulations. The use of a particular species with reference to the number therapeutic effects it can produce or the number of formulations in which it is an ingredient, is expressed as the therapeutic index and frequency index respectively. A higher index reflects a higher economic value attributable to a particular species.

- Certain species are of great importance in the treatment of particular diseases, as they happen to be the only species with that therapeutic potential, as the alkaloids of Taxus baccata in the treatment of cancer. Such species attract high market rate.
- Some species have narrow distribution and/or occur in small populations, and/or may be difficult to cultivate. Such species also command a higher price such as *Aquilaria agalocha* of Tripura, *Trichopus zeylanica* of South India, *Coptis teeta* of Arunachal Pradesh etc.
- Certain species of medicinal value like *Rauvolfia serpentina, Gloriosa superb a, Swertia chirayita, Ipomoea digitata* have been over exploited and so now occur rather scarcely in wild conditions in West Bengal.
- There are certain therapeutically active constituents produced by plants such as digoxin and digitoxin that could not be produced synthetically. So, the economic value of the plant bearing digoxin is naturally on the higher side.
- The cost involved in isolation and purification of an active principle involve several considerations. For example, it requires about a ton of leaves of Catharanthus roseus to obtain 1 gm of the alkaloid vincristine, essentially needed to treat leukemia ,Vincristine is one of the expensive plant products costing about US $ 24000/g. Vinblastine, another alkaloid from the same species, used to treat Hodgkin disease is present in quantity 1000 times more than vincristine. One gm. of vinblastine, costs about US $ 6,800. It has now become possible to convert vinblastine into vincristine through bio-transformation. Thus several factors govern the cost of the raw material and the final product of medicinal plant.

Commercial Value of Plant-based Drugs

The introduction of a single new synthetic drug into the market would take about 10 to 15 years of time and expenditure in the scale of about US $ 100 to 300 million (Abelson 1990). Plant based drugs would take a comparatively much less time and expenses than synthetic drugs. Hence, plant based medicines would be far inexpensive, unless the market prices are inflated by other considerations.

The commercial evaluation of plant based drug is different from that of the source raw material of medicinal plants, for both depend much upon the demand and the supply potential. For medicinal plants, the source material, the land use value, cost of collection/cultivation, costs of preparation, packaging and transport are the criteria. For plant-based drugs, the cost of discovery, clinical testing, processing, packaging and marketing, add up.

According to Pearce and Moran (1994) economic value of medicinal plants and plant based drugs mostly depend upon the following criteria.

The actual market value of the plants being treated.

The market value of the drugs of which the plant are the source material; and the value of the drugs in terms of their life saving potential and value of 'statistical life', which is estimated to be US $ 4 million on 1990 price.

Trade Market in Medicinal Plant Parts : The National Scenario

The markets dealing with medicinal plant parts (herbs and crude drugs) in India is a highly disorganized and less studied sector. So far little idea has been generated on the nature and quantum of trade in medicinal plant parts. The demand, supply and price structures are highly unstable. Even today, India lack a total and updated inventory on which particular plant parts are being traded in the market. There are much confusion and controversy regarding names of the items in trade. Moreover there exist practices of adulteration and also of selling totally different items as inferior substitutes. There is no definite system of quality control. There are items banned to be exported without cultivation certificate, but majority of the enforcement staff does not have any idea on identification of such items or even the existing law. The trade routes and source of most of the items are not exactly known or studied. Transit formalities and tax structure is not uniform in different states and major wholesale markets. The overall situation demand immediate attention from the Central Government, respective State Governments, and other sectors dependent on the market. A commercial database on the market has to be prepared and carrying out a detailed analytical survey on the market economy is an urgent necessity.

Trade Market in Medicinal Plant Parts : The West Bengal Scenario

In West Bengal there are two major markets dealing with herbs and crude drugs. During 2000, TRAFFIC – India conducted an all-India survey on trade in medicinal plant parts which also covered these two local markets. The following observations and findings were generated mostly during that market survey, which would perhaps give a general idea of these West Bengal two markets.

The Kolkata Market

The wholesale medicinal plant market in Kolkata is situated almost totally within Burrabazaar area in West-Central zone of the city very close to river Hooghly. The market area is roughly restricted within 1 km^2 area bounded by Biplabi Rashbehari Basu Avenue (Old Canning Street) in the South, Kali Krishna Thakur Street in the North, Rabindra Sarani (Old Chitpur Road) in the East and river Hooghly in the West. About 90% of the major trading units of Kolkata lie within this area.

The market is a relatively old one in medicinal plant trade originating in the later half of the nineteenth century. Even upto the late 1980's the market was gradually expanding. It was the most important center for medicinal plant parts collected from the North East India and from Nepal. The number of traders (including major retailers) increased slowly from around 10 to about 60, though in a considerable number of cases it resulted from split in family business. During the last 15–20 years the growth of the market has become almost stagnant; not even 10 new traders entered the market, and during the same period some also left the business. Now the number of traders from the Kolkata Market (leaving aside the minor retailers) is about

65–70 of which about 60% are wholesaler cum retailers. Only three traders from Kolkata are regular exporters. Outside these 65–70 traders there are about 15–20 traders. who in the process of dealing with Yunani medicine, deal with crude drugs and herbs. Most of these shops are located on Rabindra Sarani, south of Mahatma Gandhi Road crossing, opposite to famous Lal Masjid.

The medicinal plant trade in Kolkata is not in a flourishing state. According to most of the traders, Delhi and Mumbai have taken away the business from Kolkata. This is perhaps due to the reasons mentioned below:

- With the improvement of communication system, most of the suppliers have become direct with the major wholesalers and exporters of Delhi from where they are getting better rate/offer since today's Delhi market is a far more competitive market than that of Kolkata.
- The payment terms in Delhi and Mumbai are much better than in Kolkata. In Kolkata market generally a supplier has to wait for around 3/4 months toget their payments fully cleared and this hampers the overall trade dynamics.
- Allegedly, some of the Ayurvedic and other herbal medicine manufacturers drifted away from West Bengal for general reasons. On the contrary many new units are growing up around Delhi, Mumbai and also in South India.
- Reportedly the tax structure is much better in Delhi. It is now 4% in Kolkata but even in the beginning of 1999, it was 12%. But in Delhi the rate has been uniformly 2%. With around 10% difference in tax rate it was not feasible for the Kolkata traders to compete with the Delhi traders even through keeping lower margin. Since the lowering of the tax rate the market situation has now become relatively favourable for the Kolkata traders, but now they are facing it difficult to regain the lost business.
- The Kolkata market also suffered from loss of business from the North East India. Due to reasons beyond control and also due to legal imposition, the supply of herbs and crude drugs coming from the North East has been greatly hampered during the last 15 years. Also due

to several reasons the supply from Nepal has diverted to Delhi or in some cases to China.

- Cultivation of medicinal plants is being attempted with great effort in South India and also reportedly in West / North West India. But in the Eastern region the attempts so far made are scanty and casual. The wild stock of medicinal plants are being depleted at a very fast rate due to over exploitation or premature extraction. Unless economically viable cultivation of local/introduced species does not become successful, it is difficult for any market to flourish. Since such instances are rare, rather non-existing in West Bengal, quite naturally the market is suffering from stagnation.

In 2000, the medicinal plant part trading community of Kolkata formed an association in the name West Bengal Herb and Crude Drug Dealers' Association, to look after the interest of the community and medicinal plant trade as a whole.

Siliguri Market

The other important market dealing with medicinal plant parts in West Bengal is the market in Siliguri. There are only 4–5 wholesale traders. The market is mostly dependant on plant parts collected from North Eastern states, Sikkim and Bhutan. But recently, as mentioned earlier, due to reasons beyond control and also due to legal imposition, the supply of herbs and crude drugs coming from the North East has been greatly hampered. (What could be gathered from the secondary sources that due to the insurgency problems prevailing in the NE states, collection in the forest areas has been heavily affected. Since in most of the forests there are alleged terrorist group base, even the local people do not dare to enter the forests areas for collection of low profit items. Moreover, during the last 20 years the business community has suffered a lot from the extortionist of different political groups).

The supplies from Nepal (partially), Sikkim and Bhutan (wholly) generally enter the mainland market through Siliguri. The place also has been proximity to North Bengal forest areas. The market seems to be small but dynamic. In general, the Siliguri market is apparently a self sufficient one surviving almost solely on local supply, and feeding a significant quantity of medicinal plant parts to other parts of India. The Siliguri market is also operating in the reverse direction. Supply of several items to the retailers of North Eastern states is maintained through Siliguri, usually from Kolkata and rarely from Utter Pradesh or Delhi also.

Some General Observations on Kolkata & Siliguri Markets

The items available in the market are sold by trade names, and no authentic list/compilation of scientific names corresponding to these trade names is available in the market. Kapoor Kachri available inKolkata market might not belong to the same species which is being sold in Delhi or Mumbai Market. Lack of reference sample collection within nublic access is a big problem.

It seemed that in most of the cases even the drug manufacturers do not bother with the species level identification and use. This is apparently true also with the quality factor.

The traders are well aware about the general quality of the product they are selling and in most of the cases they give cheaper rate explaining to the buyer why the rate is cheaper (*e.g.*, slight degradation in quality due'to long time storage). In reality – apparently in majority number of cases – the traders are not mixing inferior quality product with that of superior quality. Instead they are asking for cheaper rate against quality sacrifice. In such cases actually the buyers are going for less wanted compromise.

Lack of state level inventory of medicinal plant parts in trade and also absence of an identification handbook for common people can be strongly felt. Most of the traders are not much aware of the trade restrictions and conservation laws.

Till date, the State Forest Department in West Bengal has not shown much concern over the nature or quantum of the trade in medicinal plant parts within the state.

Major Stakeholders

The major stakeholders in the field are drug manufacturers, traders in herbs and crude drugs, medicine practitioners (particularly Ayurvedic, Hakimi, Unani, Sidha, Amchi and other alternative medicine practitioners dependent mostly on medicinal plants or their derivatives), cultivators, collectors. Co-operative with cultivation and collection, Central and State Government Departments (particularly State Forest Departments and the recently formed Dept. of Indian Systems of Medicine of the Central Govt.), Scientists, research institutions, universities and NGOs.

Major Causes for the Loss of Biodiversity

General Threat to Phyto-diversity in India.

Apparently the quick decline of the vegetation cover of a country is directly linked with the rapid population growth (102 crores in India) with an ever increasing needs. The, major causes of the loss of biological diversity in India both in the hills and plains are anthropogenic. However, eight major causes for the loss of bio-diversity have been identified namely–(i) Habitat loss and fragmentation, (ii) Introduced speices, (iii) Over exploitation of plant and animal species, (iv) Pollution of Soil, Water and atmosphere, (v) Global climate change, (vi) Expansion of Industry, agriculture and forestry, (vii) Want of pollinating vectors, (viii) Unisexual flowered species.

Threat to Medicinal Plants in Particular

All the factors described above threatening the phyto-diversity in India broadly cover most of the threats to medicinal plant species. Over-exploitation and indiscriminate use of wild resources in commercial demand now-a-days playa great role not only behind the quick decline of the species concerned, but also becomes a threat for the survival of other associated species, thus acting as the major factors disturbing the entire ecosystem.

For medicinal plants, like other plant parts in commercial demand, premature exploitation is another factor which is additionally responsible for rapid destruction of wild stock. Whenever a plant is in high demand, premature exploitation is expected to threaten next season's propagation and regeneration of the species.

In this country, so far little work has been done to assess the level and extent of exploitation of plant species in commercial demand for medicinal purpose. The inventory is also not at all comprehensive or updated. Local area specific threats or species locally endangered are yet to be studied.

But going for medicinal plant cultivation in response to growing demand for certain species can be a lucrative agricultural option in West Bengal. Timely attempt to do so, would not only save some of the species from possible extinction, but also feed the state economy with assured and sustained financial benefit in near future.

Ongoing Initiatives

Protection and Conservation Strategies

Control of 'Wildlife Trade' is the most effective measure to protect the wild potential resources for their sustainable use. Using IUCN (1994) methods, Botanical Survey of India — the nodal organization responsible for monitoring and preparation of the list of plants to be included in the negative list of export — has so far listed 53 specieds for restriction in trade.

Defection of different threat groups/categories is the primary data source of taking the effective measure of protection and conservation of potential plant resources. It is roughly estimated that about 10% of higher plants are under — different threat categories. Botanical Survey of India has already published 4 volumes of Red Data Book of Indian plants (RDB) accounting relevant information of more than 800 species.

Convention of International Trade in Endangered Species of Wild flora and Fauna (CITES), is the international treaty to regulate the international trade of wild species.

The legal and policy framework include various Acts such as the Forest Act 1927; the Forest Conservation Act 1980; the Indian Wildlife Protection Act 1972, etc. These are in addition to various international treaties like CITES. Ramsar – CMS, etc. directly related to conservation of Biological Diversity, which has come into force on 29th December, 1993, which India ratified and joined on 18th February, 1994. It has been decided that for in-situ conservation, involvement of indigenous people and local ethnic communities is of utmost importance. (Brown 1990).

In Situ Conservation

The recorded forest area of India is about 77.01 million hectares but according to Landsat imagery the area covered by forests is about 64.01 million hectares. Todav India has less than 2% of total forest area in the world but supports 15% of world's population. West Bengal possesses 11% forest of its total landmass where 33% is the minimum requirement for pollution free environment. West Bengal has long history of in situ conservation and of developing in situ conservation methods at different levels, ranging from species to ecosystems.

Habitat Conservation

India has a large network of 85 National Parks and 445 sanctuaries covering about 2.5% of total land surface as well as marine ecosystem, such as Gulf of Mannar and Gulf of Cambay. It has also been proposed to increase these protected areas to 148 National Parks and 519 sanctuaries to reduce the significant gaps in various bio-geographic units. However the critical problem is not merely the conservation of a particular species or habitat, it is the continuation of the very process of evolution of all micro organisms, plants and animals in their totally as integral part of the natural ecosystem. To achieve the objective, the Government of India has designated 8 Biosphere Reserves out of 14 proposed. In West Bengal, Sunderban is one of the Biosphere Reserves. World Heritage Conventionn has designated 5 natural sites as world Heritage sites. Sunderbans of West Bengal is one of the Heritage sites.

A National Committee on wetlands, Mangrove and Coral Reefs has identified 21 wetlands, 15 mangrove areas and 4 coral reef zones for conservation and scientific management.

Ex-situ Conservation

To complement the in situ conservation, considerable attention has been paid to ex situ methods of conservation. The collection (from within and outside the country), preservation, multiplication and supply of genetic resources (for research only) is done through National Bureau of Plant genetic Resources, New Delhi for wild relatives of crop plants. In addition to 66 Botanic Gardens (including 33 university botanic gardens), the Department of Biotechnology has initiated germplasm facilities. India is the overall coordinator for the establishment of Gene Bank of Medicinal and Aromatic Plants among 6–15 countries.

For large scale multiplication of forest tree species two tissue culture pilot plants have been established at National Chemical Laboratory, Pune and at Tata Energy Research Institute, New

Delhi. Plant Issue Culture laboratories have also been set up in different National Institute.

The West Bengal Scenario

So far, no major conservation measures aiming at conservation of medicinal plants in particular or ensuring sustainable use of the same has been taken in West Bengal. Cultivation of some commercially valuable species has been attempted in some areas mostly by some private entrepreneurs with the motive of searching for lucrative agricultural options. Some research institutions like Bidhan Chandra Krishi Viswavidyalaya has attempted to develop agro-techniques for certain species. Only recently, the state government has started showing interest in the field. It has now become evident that going for medicinal plant cultivation in response to rising demand for certain species can be a lucrative agricultural option in West Bengal. Timely attempt to do so, would not only save some of the species from possible extinction, but also feed the state economy with assured and sustained financial benefit in near future.

Proposed List of Cultivable Medicinal Plants in West Bengal

Out of about 4000 medicinal plants in India, West Bengal possesses about 700 species covering aromatic plants, species and herbal vegetables of which about 75 species are known to be commercially operated either collected from natural sources or from cultivation. Some of them are supposed to be agri-alternative in West Bengal and commercially viable in cultivation.

Suggested Measures and Action Plan

Some suggested measures and action plan which should be taken up to ensure proper sustainable use of medicinal plant parts in India in brief are as follow (the expected actors and required time period is shown within parenthesis against each suggested action):

1. Preparation of a state level inventory of medicinal plant parts in trade in the major markets within the state. Species level scientific identification is extremely necessary in case of each items. Basic minimum

information which are required to be collected on the products are: (i) Trade or local names/Scientific or botanical name. (ii) Source : Cultivated/Wild/Forest/non-forest: (iii) Availability : (Districtwise and forest area-wise in case of forest products). (iv) Known/claimed medicinal properties (v) Current price range. (vi) Demand and availability trend through last five years. (vii) Anticipated conservation status. (viii) Cultivation status within the state. (ix) Agro-techniques (if known). (x) Availability in other parts of India (statewise and major trade centers) (Suggested Actors: Government local NGOs, Universities or other research institutions. Time period : 1 year). Thorough literature survey - areawise (ecosystemwise) and groupwise up to species level and documentation. (Suggested Actors: Government, NGOs, Universities or other research institutions. Time period: 1 year).

2. Rapid documentation of empiric knowledge base of tribal and ethnic communities and rural people of India about medicinal plants and their uses. (Suggested Actors: Government, NGOs, Universities or other research institutions. Time Period : 3 years).
3. Rapid documentation of traditional conservation practices existing among the tribal, ethniccommunities and rural people of West Bengal. (Suggested Actors: Government, NGOs, Universities or other research institutions. Time Period : 2 years).
4. Identification of threatened habitat and taxon (Suggested Actors : Government NGOs. Universities or other research institutions. Time Period: 2 years).
5. Preparation of State Herbarium of medicinal plants occurring in India (Suggested Actors : Government, NGOs, Universities or other research institutions. Time Period : 2 years).
6. Preparation of a reference sample collection of medicinal plant parts in trade for the country allowing public access to the collection. (Suggested Actors : Government, NGOs, Universities or other research institutions. Time Period : 2 years).

7. Preparation of a pictorial identification handbook on medicinal plant parts in trade. (Suggested Actors : Government, NGOs, Universities or other research institutions. Time Period : 1 year).
8. Formation of a marketing and development board in West Bengal for medicinal and aromatic plants and phyto pharmaceuticals. Such a Board can interact with growers and user industry to bring stability in their production. demand, price, quality and can also help in fostering international trade. The Board should invite members and representatives from all segments of stakeholders particularly ayurvedic drug manufacturers, herbs and crude drug traders. medical practitioners and botanists. (Suggested Actor : State Government Time Period : Immediate).
9. State level initiatives to introduce and promote cultivation of medicinal plants occurring naturally within the state which are in high demand. (Suggested Actors : Govt., NGOs. Time Period : 5 years).
10. Promoting cultivation throughout the state giving priority to items which can he exnorted with cultivation certificate authentication only. (Suggested Actors : Govt., NGOs Time Period : 5 years).
11. Developing agro-technique for as many as possible plants. So far agro-technique could be developed for 42 medicinal plants only. (Suggested Actors: Government, NGOs, Universities or other research institutions. Time Period: On-going process, Long Term).
12. Providing scientific identification facilities to public at nominal cost. The facilities should be extended to assess also the quality of the products. Suggested Actors: Government, NGOs, Universities or other research institutions. Time Period: Should be started within 1 year, followed by possible expansion).
13. Initiating scientific research on the alleged medicinal properties of each of the plant parts and anticipated side effects. (Suggested Actors: Central and the State Government, Research Institutions, R&D section of major drug manufacturers, Universities, and NGOs. Time Period : Long Term).

14. Ensuring uniformity in transit formalities for medicinal plant parts with other Indian states. (Suggested Actors : State Government, Central Govt. Time Period: Immediate).
15. Ensuring uniformity in tax rate for medicinal plant parts with other major markets in India, particularly at par with the New Delhi market. (Suggested Actor: State Government Time Period: Immediate.)
16. Preparation of specific list of banned medicinal plant items for West Bengal market. (Suggested Actor: State Government Time Period : Immediate.)
17. Training enforcement staff regarding existing laws on medicinal plants and identification of restricted species. (Suggested Actor : Central and State Government, NGOs. Time Period : Immediate.)
18. Preparation of single specific list containing items taxable as medicinal plant parts for the country. For taxation, it is essential to include each plant part in one list only, *i.e.*, if any item has been included in the list of dried fruits, it should never be considered as medicinal plant part, and vice versa. Definition under one list is necessary and less confusing in case of any legal settlement. (Suggested Actor: State Government Time Period: Immediate).
19. Publication of a manual regarding laws dealing with medicinal plants, inventory list of herbs and crude drugs in West Bengal market, banned item list, collection & transit formalities within the state, list of registered whosesale traders/exporters/manufacturers and research laboratories equipped to deal with medicinal plants. (Suggested Actors: Government, NGOs. Time Period: 3 years).
20. Establishing infrastructural facilities for ex-situ conservation especially seed bank and medicinal plant nursery. (Suggested Actors: Government NGOs, other research institutions. Time Period : Should be started within 1 year, followed by possible expansion).

Chapter Three
Ayurveda : An Indian System

This elaborate subject is not being discussed in this Chapter. Enormous amount of Literature exists on the subject. Legendary figures in Ayurvedic medicines are Indian Physicians *e.g.*, Atreya, Mahabhorati, Nagarjun Sagar, Vagabhata, Sushruta and others.

The period between 800 B.C. and 1000 A.D. could be considerd as the golden age of Indian System of Medicine.

"Ayurvedic medicine", according to Encyclopedia Britanica "is an example of a well-organized system of traditional health care, both preventive and curative".

Some useful plants, plant and some other recent information have been presented in this Chapter.

Of 84 Ayurvedic plant species recorded in a table, only 20 species have been recorded as safe, 28 are cultivated, 16 as sporadic. The depleted genera are Artenesia, Asparagus, Cessampelos, Cordia, Curculiga, Gymnema, Hemidesmus, Nardostachys, Nelumbo, Paederia and Rauvolfia.

The indication seems to be bad.

Ayurveda

The backround history of Ayurveda has its origin about 4500 B.C. ' Charaka Samhita' and Susharata Samhita are considered to be most vital documents of Ayurveda.

The evolutionary history and various stages of development are not the subject of discussion in this treatise. As such only broad outline on Ayurveda System is discussed.

Ayurveda–An Indian System

Enormous amount of literature exists on this subject. During the last sixty years voluminous work has been done on various facts of Ayurveda and on the drugs used in India. The author does not find it necessary, in the present context, to rewrite the history starting from Vedic Period since 4500 B.C., records of which exist.

A brief recapitulation, however, may be relevent which are –

- Most vital documents were compiled by the great Charaka and Sushruto in Charaka-Samhita' and Sushrata-Samhita'.
- In the seventh century A.D.' Astango Hridaya Samhita' give a wide range of information.
- The period between 800 B.C.' and 1000 A.D. is considered as golden age on indian System of medicine.
- The Indian Ayurvedic system of medicines and treatment suffered a stagnation due to prolonged Muslim and the British rule.
- In 1563 Garcia D'orta a portugeese physician a tretise on indian medicine.
- The work of several foreigners on Indian Drugs are commendable. Mention may be made of the names of the following:
  - Flemming (1910), Ainsic (1913), Roxburgh (1834), Waring (1868), Mohidin Sheriff (1869), Warden and Hooper (1890-93), Georage Wall (1889–1904).
  - The Scientists of CSIR are engaged in updating various technologies particularly agro and processing technology resulting in production of derivations like phytochemicals, essential oils, Oleo resins etc.
  - The famous Scientist Dr. (Mrs.) Asima chatterjee thinks the therapeutic use of Indian medicine dates back to 4500–1600 B.C *i.e.*, early are of Rig-Veda. This led to evaluation (2500–600 B.C.) of Ayurveda which literally means "Science of Life."
  - Systematic and concerted Scientific research was pioneered in India on indigenous plants by Dr. R.N. Chopra and his colleagues in the school of Tropical Medicine, Kolkata.
  - Encyclopaedia Britannica (Marco, 23,906,1988) writes "Ayurvedic medicine is an example of well-oraganized system of traditional health care, both preventive and

system of traditional health care, both preventive and curative. It is still a form of health care, both preventive and curative. It is still a form of health care in large part of Eastern World especially in India, where a large number of population use this system exclusively or combined with modern midicines."

- Dr. K.P. Biswas and A. Ghosh's work on medicinal plants has been recoded in the books entitled. " Bharatiya Bonousodhi". This work gives a wealth of information. The work of Nagendra Nath Sen Gupta , K.M. Nadkarni and Dr. Sibakali Bhattacharjee present an indepth information in various work done in medicinal plants of India.

Nadkarni in his book "Indian Materia Medica" describes remedial properties of about 550 plant species. Of these, 90 are tree species, 28 are climbing species, 35 are shrub species and 397 are herb species. Dr. Bhattacharjee in books' Chiranjib Banousodhi' has discussed plant properties of more than 800 species and has listed more than 3000 local names of medicinal plants of various states of the country.

On the context of these valuable work, it is worthwhile to ascertain the present status of medicinal plants to find out how far this wealth needs protection.

Originated in the Vedic times around 5000 years ago. Ayurvedic formulations which are an ancient health system; use a combination of, selection of around 1200 species about 500 of which are commercially traded. Ayurveda uses medicinal plants in various forms, some of which can be gathered only by destructive harvesting: in 30 per cent cases only the roots are used, in another 13 per cent only the bark and it is only in about 16 per cent that the whole plant is used in other cases, medicines use the fruits, leaves, flowers, rhizome seeds etc. It is commonly thought that medicinal plants are mainly herbs, but in fact about one-third are trees—this has implications for conservation and management of supplies to the industry. The majority of plants used in ayurveda are procured from the wild, though around 10 per cent are cultivated on private lands.

Trades on medicinal plants of India may be described as extremely complex, secretive. traditional, badly oraganized, highly under-estimated and unregulated. There is no macro level information available for assessing the nature and full extent of the trade; there are only 'guesstimates' based on local inventories and micro studies. Identification of species and volumes traded is futher complicated by the fact that there is no reliable correlation between trade names and botanical names, and names used for particular species may change along the supply chain. Conversely, the same trade name is at times used for several species, especially if they are used for similar purposes. Hence, for the purposes of this study, twelve of the most representative species were selected for detailed research into the conservation, collection, cultivation and trade of medicinal plants. These are *aloe vera, Chlorophytum borivillianum, Commiphora, wightii, Embelia ribes, Embillica officinalis, Nardostachys grandiflora, Picorrhiza kurroa, Rauwolfia serpentina, Saraca indica, Swertia chirata, Terminalia chebula and Withania somnifera.*

Trades on medicinal plants are localed in 6 major, 21 medium and 37 minor markets spread across the country. The major centers, located at the heads of the routes taken by the medicinal plants, are big cities including the four metros. Major export take place from Delhi, Mumbai, Chennai and Tuticorin. In terms of total volume of the 12 species traded in 1999–2000, Mumbai tops the list with about 3,300 tonnes, followed by Delhi with about 2,000 tonnes. The survey indintified a number of facators that affect the final price. Volumes traded are directly proportional to the prices of the raw material, which in turn are proportional to the abundance/availability of the species. There is also a connection between the part used and prices, so that species that are distructively harvested seem to be more expensive. High altitude species such as *Nardostachys grandiflora, Picrorrhiza kurroa, Swertia chirata* also are high value species. Price also increases with the distance of the source of raw material from the market.

Use of Medicinal Plants in the Ayurvedic Industry

The prescribed reference texts of Ayurveda namely Charak, Sushruta, and Bagbhatta describe use of 1,100, 1,270 and 1,150 medicinal plants respectively in drug formulations (Gupta 1993.) Annex 4 lists some of the most important medicinal plants used in ayurveda and the conditions they are used to treat. Ayurveda uses medicinal plants in various forms: fruits, leaves, flowers, rhizome, bark, roots and seeds, etc. (see Table 3.1). Sometimes whole plants are used but mostly, some parts are used to prepare the formulations. Estimates suggest that about 16.5 per cent of the usage requires whole medicinal plants while in the rest one or a combination of plant parts is used (FRLGT 1997).

Table 3.1: Analysis of Plant Parts Used in Ayurvedic Industry

Parts used	Percentage
Roots	29.6
Rhizome	4.0
Leaves	5.8
Flowers	5.2
Fruits	10.3
Seed	6.6
Stems	5.5
Bark	13.5
Wood	2.8
Whole plant	16.5

(Source : FRLHT 1997)

India harbours a wide range of medicinal and aromatic plants mostly sued in Ayurvedic, Unani. Siddha, Homepathic, Allopathic and other alternate medicinal practices such as folk medicine, home remedies, household remedies naturopathy, tantra-theraphy, Amchi and tribal medicine. The plants used in alternate medicine are awaiting a touch of modern knowledge.

Out of 17,500 flowering plants species known from India, more than 4000 species are used as medicinal plants of which 300 species yield gum and dyes and about 100 species yield essential oils and are used as raw material in drug industry. About 200 drugs are of animal and mineral origion.

Keeping in mind this perspective it can be mentioned that the rigveda (4500–1600B.C) an old reposlitory of human knowledge, quoted 67 plants of medicinal value. The Atharva Veda and the Yajurveda mention 200 and 81 species of plants of the therapeutic value. Charak (1000 B.C.) and Sushruta (800 B.C.) quoted 341 and 395 medicinal plants respectively in their Samhitas. Ethno-botanical studies have added about 1680 species of medicinal value in the existing list of medicinal plants in India.

In last 75 years the basic active principles like salicylic acid, alkaloids, saponins, glycosides, steroid compounds have been isolated from higher group of plant except Ephidrin from from Ephedra (gymnosperm), Taxol from Taxus (gymosperm), new alkaloid from Lycopodium, ergot from civiceps fungus and other antibiotic from fungi groups. During 1950, only 2000 alkaloid was known to the world. By 1970, the number increased to 4000, and in 2000 A.D. it has been estimated to be about 10,000 alkaloids, of which 5000 are chemically less known.

The discovery of hallucinogenic use of morning glory has stimulated future studies of Datura, Canabis, and Paper in India. The hallucinogenic drugs are very much in demand in today's world. Very recently steroidal sapogenious substance has also attracted phyto-chemists to give more attention in their research.

Herbal Medicines

The vidic and post-vedic periods since 3500 B.C. saw some celebrated Indian physicians *e.g.*, Atreya, Mahabarati, Nagarjun Sagar, Vaghbhata, Sushruta, Charaka, who were legendry fighers of Indian medicine.

Although the tradiditional Medicines bad been repleced to a great extent by chemical and synthetic medicines, herbal medicines are once again gaining ground das the synthetic one are costly and found toxic and proved to be health hazardous.

People suffering from chronic diseases are willing to use trational medicines that the synthetic ones. Several countries also are looking back towards the " Traditional folklore Medicines". China resisted all imports of modern medicines by use of more and more traditional ones. So herbal wealth of India needs concservation so that traditional species could be sused more and more.

In 1948, Col. R.N. Chopra submitted a report recommending revival of traditional Indian Medicine and its synthesis with modern medicines. Such integration were also advocated by the government of India.

Revival of Cultivation of medicinal plants— In recent years several states and private organizations have start cultivation and conservation of medicinal plants in garden and Arbourata. One such endeavour adopted by Orissa State (The Statesman new Service, Feb.6,2001) is worth mentioning.

The first phase of an ambitious " Medicinal garden" was launched here recently with the District Enviroment Society, Khurda, making a modest beginning by plating 50 defferent species of medicinal plants near patrapara village.

The species planted in the first phase area of 40 acres have been procured from the nearby Brebera forests and from the silviculturist, Orissa.

Funded by the forest department, district rural development agency and by a Rajya Sabha MP Mr. BJ Panda, the plan is to develop Medicinal garden over a total area of 250 acres for plantation in three phases. The area happens to be a degraded Sal forest and is close to the state capital forest department officials said.

Emphasizing on the need to recognize, preserve and develop such plants, they said 8000 species of medicinal plants are used by different medicine systems in the country.

The World Health Organisation (WHO) has estimated that 80% of the population of developing countries on traditional medicine mostly derive from plants for their primary healthcare needs. The demand of medicinal plants is increasing throughout the world. 90% of the drugs used in Indian Systems of Medicine & Homoeopathy (ISM&H) are plant based and collected from wild sources.

Government of Indian has set-up a national level body called the Medicinal Plants Board in November 2000 for the development and sustainable use of medicinal plants in the country. The Board aims at making the cultivation of medicinal plants and its sustainable management besides co-ordinating all matters related to medicinal plants and to make it a people's movement. The medicinal plant species included in this publication represent high demand plants, which can give good remuneration if a tie up with buyers is arranged.

The Board will undertake the following activities:

1. Promote encouragement for cultivation of selected medicinal plants backed by buy-back arrangement.
2. Enocurage States and UTs to registering raw drug traders and cultivators so that source of supply of medicinal plant is monitored as a measure to promote quality control, safety and efficacy of drugs.

3. Facilitate measures, which enhance efficency, cost effectiveness and upgradation of harvesting, dryin, grading packaging, transportation and storage of medicinal plants.

4. The following thirty-one (31) species, which are in high demand both in domestic and international markets are to be brought into cultivation status as these constitute a bulk of the ingredients used in the preparation of ISM&H and herbal products. This list will naturally undergo changes from time to time.

Table 3.2

Sl. No.	Common Name	Botanical Name	
1.	Amla	*Emblica officinalis* Gaertn	Perennial tree
2.	Ashok	*Saraca asoca(Roxb.)* de Wilde	Perennial tree
3.	Ashwagandha	*Withania somnifera (Linn.)* Dunal	Annual herb
4.	Atees	*Aconitum heterophyllum* Wall.ex Royale	Biannual herb
5.	Bael	*Aegle marmelos (Linn) Corr.*	Perennial tree
6.	Bhumi amlaki	*Phyllanthus amarus Schum &Thonn.* (p.niruri Linn.)	Annjual herb
7.	Brahmi	*Bacopa monnieri (L.) pennell*	Annual herb
8.	Chandan	*Santalum album* Linn.	Perennial tree
9.	Chirata	*Swertia chirata* Buch-Ham	Biannual herb
10.	Giloe	*Tinospora cordifolia* Miers.	Perennial climber
11.	Gudmar	*Gymnema sylvestre* R. Br.	Perennial climber
12.	Guggal	*Commiphora wightii (Arn.) Bhandari*	Annual climber
13.	Isabgol	*Plantago ovata* Forsk.	Annual herb
14.	Jatamansi	*Nardostachys jatamansi* DC	Perennial herb
15.	Kalihari	*Gloriosa superba* Linn.	Annual climber
16.	Kalmegh	*Andrographis paniculata* Wall. Ex Nees	Annual herb
17.	Kokum	*Garcinia indica Chois*	Perennial
18.	Kuth	*Saussurea costus* C.B.Clarke *(S.lappa)*	Annual herd
19.	Kutki	*Picrorhiza kurroa* Benth ex Royle	Annual herb
20.	Makoy	*Solanum nigrum* Linn.	Annual herb
21.	Mulethi	*Glycyrrhiza glabra* Linn.	Perennial herb

contd...

Table 3.2 – *contd...*

Sl. No.	Common Name	Botanical Name	
22.	Musali Safaid	*Chlorophytum arundinaceum* Baker *(C. borivillianum)*	Annual herb
23.	Pashan Bheda (Coleus)	*Coleus barbatus* Benth.	Annual herb
24.	Pippal	*Piper longum* Linn.	Parennial climber
25.	Rasaut (Daruhaldi)	*Berberis aristata* DC.	Parennial shrub
26.	Sarpgandha	*Rauwolfia serpentina* Benth. ex Kurz	Perennial herb
27.	Senna	*Cassia angustifolia vahi.*	Under shrub
28.	Shatavari	*Asparagus racemosus willd.*	perennial climber
29.	Tulsi	*Ocimum sanctum* Linn.	Annual herb
30.	Vai Vidang	*Embelia ribes* Burm. f.	Perennial shrub
31.	Vatsnabh	*Aconitum ferox* wall.	Perennial herb

5. Under general and specialized surveys of the national and international market for medicinal plants and products for identifying niche areas.
6. Motivate and encourage States/Uts to set up State Medicinal Plants Board/Vanaspati Van Societies who can give a focus and direction to medicinal plants related activities.
7. Support manufacturers/NGOs and representative individuals for participation in international fairs, seminars and meetings with a view to create awareness and explore the international market for plant based herbal products.
8. Support R & D studies in the areas of post harvest management including increasing shelf-life introducing better storage techniques and agro-techniques, enhnce bio-availability to be taken up through CSIR, NBRI, CIMAP, ICFRE, RRLs, DBT, Horticulture and Forest Departments.
9. Launch efforts to create mass awareness about the importance of medicinal plants in all strata of society, rural and urban.

India is bestowed with a treasure of medicinal plants. The supply base of 90% herbal raw drugs used in the manufacture of

Ayurveda, Siddha, Unani & Homeopathy systems of medicine is largely from the wild. Besides this, plant are also used in various industries proudcing herbal items other than medicines. This wild source is speedily shrinking day-by-day. Therefore, there is a need for conservation and sustainable use of medicinal plants. Cultivation is clerly a sustainable alternative to the present collection of medicinal plants from the wild. This can be a potential provider of returns to the farmers/cultivators.

Keeping the above concept in view, the Department of Indian Systems of Medicine & Homoeopathy has indentified 31 (thirty-one) potential medicinal plants. In the present booklet brief cultivation practices together with relevant information on these medicinal plants have been presented for the interested growers/ cultivators either as a single crop or for intercropping.

(Source : Modified by RBs Rawat, National Medicinal Plant Board)

Table 3.3 : Important Medicinal Plants Used in Ayurveda

Plant	Sanskrit name	Main indications/effects	Status
Acorus calamus	Vacha	Nervine, antispasmodic, sedative, stomachic, expectorant, laxative, diuretic,	
Artemisia absinthium	Indhana	Anthelmintic	D
Artemisia vulgaris	Nagadamni	Anthelmintic, expectorant	D
Asparagus racemosus	Shatavari	Antispasmodic, antidirrhetic, demulcentPS	
Azadirachta indica	Neem	Skin disease, antibacterial	S
Bacopa monnieri	Brahmi	Nervine tonic, diuretic,sedative	S
Boerhavia diffusa	Purnarnava	Diuretic, expectorant, laxative	
Boswellia serrata	Shallaki	Antiarthritic, analgesic, antiinflamatory	PS
Buchanania lazan	Plyala	Skin disease, laxative	PS
Butea monosperma	Palasa	Diarrhea, flatulence, anthelmintic	PS
Callicarpa macrophylla	Pringu	Joint pain, skin disease, blood disease	S
Calotropis gigantea	Alarka	Bronchitis, diarrhoea, cancer	PS
Cannabis indica	Bhang	Insomnia, dysmenorrhea	PS
Capsicum annum	Katuvira	Rubefacient, stimulant	C
Capicum carvi	Krishnajira	Flatulence, stomachic	C

contd.....

Table 3.3 – *contd...*

Plant	Sanskrit name	Main indications/effects	Status
Carum copticum	Ajwayan	Spastic bowel, flatulence, dyspepsia	C
Cassia angustifolia	Markandika	Constipation, liver disease, joint pain	
Cassia fistula	Argbhada	Ringworm, constipation, fever, antibacterial	PS
Cedrus deodara	Devadaru	Fever, diarrhea, urinary disease	PS
Centella asiatica	Manduk-parni	Sedative, alterative, anxiolytic	S
Cichorium intybus	Kasni	Emmenogogue, digestive	S
Cinnamomum camphora	Karpoor	Diarrhea, nervousness, muscular pain fever	C
Crocus sativus	Kumkuma	Nervine sedative, emmenogogue, aphrodisiac	S
Cinnamomum zeylanicum	Twak	Dyspepsia, flatulence, diarrhea, menorrhagia.	C
Cissampelos pareira	Laghu Patha	Spastic bowel, uterine prolapse, alterative	D
Clitoria ternatea	Aparajita	Cedema, anthelmintic, demulcent	C/PS
Cocus nucifera	Narikela	Fever, pharyngitis, skin disorder, alterative	C/PS
Coleus aromaticus	Pashan-bheda	Kidnet stones, conjunctivitis, spastic colon	C
Cordia obliqua	Shelsh-mantaka	Expectorant, colic, dyspepsia, ulcers, cough	D
Coriandurm sativum	Dhanyaka	Flatulence, colic, joint pain, antiseptic	C
Crinum deflexum	Sudershan	Emetic, inflammatory conditions	C
Cuminum cyminum	Jeeraka	Diarrhea, dyspepsia, antiseptic, hookworm	C
Curculigo orchioldes	Talamulika	Hemmorroids, asthma, kidney stones, skin	D
Curcuma longa	Haridra	Arthritic pain, anti-inflammatory, skin disease	C
Curcuma zedoaria	Shati	Cough, asthma, leukorrhea, tonnesillitis	C/D
Cynodon dactylon	Doorwa	Diuretic, styptic,hematuria, hemorrhoids	PS
Cyperus rotundus	Mustaka	Anti-inflammatory, flatulence, fever, estrogenic	PS
Datura metal	Daturra	Anti-spasmodic, joint pain, asthma, dysmenorrhea	S

contd.....

Table 3.3 – *contd...*

Plant	Sanskrit name	Main indications/effects	Status
Daucus carota	Garijara	Blood purifier, nervine tonic, jaundice	C
Dolichos biflorus	Kulitha	Edama, kidney stone, asthma, dysmenorrhea, tumours	C
Eclipta Alba	Bhringaraj	Hapatic, deobstruent and tonic, alliterative, emetic purgative, antiseptic, antiviral	S
Elettaria cardamomum	Elaichi Chhoti	Bronchitis, flatulence, dyspepsia, hemorrhoids	C
Emblica officinalis	Amalaki	Fruit: cooling, laxative, stomachic, tonic, diurentic	PS/C
Evovulus alsinoides	Shankha-pushpi	Anxiety, diarrhoea, bronchitis, memory loss, fever	S
Ferula foetida	Hingu	Glatulence, cough, constipation, palpitation, aphrodisiac	C
Ficus religiosa	Aswatha	Ulcers, skin disease, diabetes, constipation	PS
Ficus racemosa	Udambara	Diarrhoea, hemorrhoids, bleeding, disproders, antiseptic	PS
Foeniculum vulgare	Satupuspa	Cough, flatulence, dysmenorrhea, hookworm edema	C
Grewia hirsuta	Nagbala	Diarrhoea, wounds, heart disease, fever	S
Gmelina arborea	Gambhari	General tonic, to increase strength, antiviral, indigestion	PS/C
Gymnema sylvestre	Meshas-tringa	Diuretic, astringent, hypoglycemic, regerant, stomachic	D
Hemidesmus indicus	Sariva	Excellent alternative, to increase appetite, cough, skin	D
Holarrhena antidysenterica	Kutaja	Diarrhoea, dysentery, amebiasis, anthelminthic	PS
Hyoscyamus niger	Yavani	Chronic dementia, hysteria, palpitations, asthma, sedative	C
Hyssopus officinalis	Zupha	Cough, asthma bronchitis, amenorrhea	C
Ipomoea digitata	Vidari	Cough, hoarseness, respiratory stimulant, tonic	S
Justicia adhatoda	Vasaka	Bronchitis, asthma, jaundice, antispasmodic	Ps
Linum usitatissimum	Uma`	Cystitis, bronchitis, boils, anti-torant, demulcent	C

contd.....

Table 3.3 – *contd...*

Plant	Sanskrit name	Main indications/effects	Status
Luffa acutangula	Koshataki	Splenomegaly, emetic, skin disease, expectorant	C
Madhuca longifolia	Madhuca	Tonnensillitis, cough, rheumatic joints, diabetes, appetizer	PS
Michelia champaca	Champaka	Gastritis, chronic arthritis, emmenagogue, diuretic, colic	S
Mimosa pudica	Lajjalu	Menorheagia, hemorrhoids, skin wounds, diarrhea	S
Mimusops elengi	Bakula	Tonic, cardiotonic, urogenital disease, snakebite, skin sores	C
Morinda citrofolia	Ach	Acne, eczeme, hyperlipidemia, bronchitis, diarrhoea	S
Moringa oleifera	Sigru	Source of Vitamin c, colds, boils, fever, joint pain, gout	PS/C
Mucuna pruriens	Kapika-chchha	Nervine tonic, aphrodisiac, Parkinsonism, hypercholesterolemia	D
Narostachys jatamansi	Jatamansi	Nervousness, anxiety, dysmenorrhea, insomnia, hair tonic	D
Nelumbo nucifera	Parijata	Refrigerant, sedative, demulcent	D
Nyctanthes arbortristis	Parijata	Liver diseases, constipation anthelminthic, anthistaminic	C/PS
Ocimum sanctum	Tualsi	Demulcent, expectorant, anticatrrhal antispasmodic, anthelminthic	C
Paederia foetida	Prasarni	Rheumatic joint pain, edema, bladder stones, inflammation	D
Papaver somiferum	Ahiphenam	Anxiety, diarrhoea, aphrodisiac, sedative	C
Peucedanum graveolens	Satapushpi	Flatulence, colic, abscesses, digestive	C
Phyllanthus fraternus	Bhumia-malaki	Jaundice, liver disease, fever, genitourinary disease, edema	S
Picrorrhiza kurroa	Katuki	Hepatitis, asthma, anorexia	C
Piper nigrum	Maricha	Dyspepsia cough, pharyngitis, headache, diarrhea	C
Plantago ovata	Isaphgol	Constipation, colitis, irritable bowel, cystitis	C
Plumbago zeylanica	Chitraka	Abortifacient, warts, rheumatic joint pain	C
Premna integrifolia	Agnimantha	Flatulence, fever, arthritis, liver deobstruent	S
Prunus amygdalus	Badama	Mental energy, general tonic esp. nerve and kidney, semen	S

contd.....

Table 3.3 – *contd...*

Plant	Sanskrit name	Main indications/effects	Status
Pterocarpus sanatalinus	Rakta Chandan	Skin tonic, liver disorders, fever	C/D
Punica granatum	Dadima	Anthelminthic (esp. tapeworm), diarrhoea, dyspepsia	C
Randia ducmetorum	Madana	Fruit and rind are emetic, diaphoretic and antispasodic, bark is sedative and nervine calamative	S
Rauvolfia serpentina	Sarpa-gandha	Hypertension, anxiety, insomnia, colic	D

Ayurvedic Formulations

Ayruvedic medicines can be classified as ayurvedic classical formulations and patent and proprietary formulations. The ayurvedic classical formulations include those medicines that are menufactured according to prescriptions given in one of the ancient ayurvedic texts, while the patent and proprietary medicines are the outcome of research and development efforts of manufacturing companies. The concept of combination of ingredients and hence formulations in ayurveda. Table 3.4 indicates the extensive range of formulations in which particular species may occur.

Table 3.4 : Frequency of Occurrence of Medicinal Plants in Herbal Formulations in India

Common name	Botanical name	No. of herbal formulation	Status
Harra	*Terminalia chebula*	Herra, behera and aonla	PS
Behera	*Terminalia belerica*	together used in 219	PS
Aonla	*Emblica officinalis*	Formulations	PS
Yashtimadhu	*Glycyrrhiza glabra*	141	C
Pipali	*Piper longum*	125	C
Vasaka	*Adhatoda vasica*	110	PS
Ashwagandha	*Withania somnifera*	109	C
Mastak (Motha)	*Cyperus rotundus*	102	PS
Guduchi	*Tinispora cordifolia*	88	D
Daruharidra	*Berberis aristata*	65	D
Gokshura	*Tribulus terestris*	65	D
Kutja	*Holarrhena antidysenterica*	59	PS
Punarnava	*Boerhavia diffusa*	52	S

Source: Biotech Consortium India Ltd. 1996

Background to the Unani and Siddha System of Medicine

Unani System

The unani system of medicine was introduced in India by Arabs and Persians from Greece where it originated between 460 BC–377 BC. According to Hippocrates, disease is a normal process and its symptoms the reaction of the body to the disease. The body has four humours namely Blood (Dam), Phlegm (Balgam), Yellow bile (Safra) and Black bile that keep the equilibrium. The humours have specific temperament and the temperament of a person is expressed as being sanguine, phlegmatic, choleric and melancholic according to their level in the body. The System believes in the presence of some self-preservation mechanism in human body. The diagnosis and treatment are based on the concept of temperament and changes in temperament are related to changes in the balance of humours. Drugs are made of herbal, animal and mineral origin. The drugs stimulate and strengthen the defence mechanism and normalize the imbalance.

Table 3.5 : Top 17 medicinal plants consumed by Baidyanath, Jhansi

	Trade name	Botanical name	Status
1	Amla*	*Emblica officinalis*	PS
2	Ashok*	*Saraca indica*	C
3	Babul	*Acacia arabica*	PS
4	Ghee kunwar*	*Aloe vera*	C
5	Urad	*Phaseolus mungo*	PS
6	Harra* (Large and Small)	*Terminalia chebula*	PS
7	Munkka		
8	Arjun	*Terminalia arjuna*	PS
9	Adusa	*Adathoda vasica*	PS
10	Baheda	*Terminalia belerica*	PS
11	Guduchi	*Tinospora cordifolia*	D
12	Kateri (Large and Small)	*Solanum surattense*	S
13	Rasna	*Pluchea lanceolata/ Polygonum spp./ Dendrophthe facultata*	S
14	Shankhpushpi	*Convolvulus pluricaulis*	D
15	Jawasa	*Alhaqi maurorum/ pseudalhaqi*	D
16	Ashwagandha*	*Withania somnifera*	D/C
17	Safed musli*	*Chlorophytum borivillianun*	D

*Species selected for detailed market study

Siddha System

Siddha means achievment of perfection, saintly figures who achieve excellence through the practice of yoga, promoted the system in Tamil Nadu. The manuscripts are in Tamil. It is believed that eighteen siddhas contributed towards the development of siddha medicine. It is largely therapeutic in nature.

The principles and doctrine of this system are similar to that of ayurveda. The difference is basically linguistic. According to this system, the human body is the replica of the universe, as are the food and drugs, irrespective of their origin. This system also accepts the five-element theory and the tridosha theory as in ayurveda. The diagnosis involves identifying causes by examining pulse, urine, eyes, voice, body colour, tongue and the state of the digestive system (ITCOT 1999).

The study was divided into three phases:

Phase I: An overview of the ayurvedic industry in India, especially in relation to its consumption of raw material (*i.e.*, medicinal plants).

Phase II: Study of pricing, value addition and information flow mechanisms at different points in the supply chain ranging from the primary collector/cultivator to the processing industry.

Phase III: Investigation of the potential application of market based instruments to ensure both supply of sustainably managed medicinal plants to the ayurvedic industry and better returns for primary collectors/ cultivators.

Chapter Four

Potential Drug Plants of Undivided India Listed by Kirtikar and Basu

The list of drug plants prepared by Kirtikar and Basu covers large area of undivided India. In marking the 'Status' of various species the author has not corrected any nomenclature of species. The 'Status' column shows making of 'dash' (– sign), which mean not known. The status column shows preponderance of 'Sporadic' species. The legend used in all the lists is as follows:

S = Sporadic.

S* = Sporadic with concentrated patches of regeneration.

T = Threatened (also D = Depleted).

PS = Presently safe.

C = Cultivated.

Table 4.1 : List of Medicinal Plants (Kirtikar and Basu)

Scientific Name	Status
Ranunculaceae	
Clematis	
napaulensis	T
triloba	T
smilacifolia	S
gouriana	S
graveolens	–
Anemone	
obtusiloba	T
Thalictrum	
foliolosum	T
Adonis	
aestivalis	–
Nigella	
sativa	C
Ranunculus	
trichophyllus	–
lingua	–
sceleratus	PS
pensylvanicus	–
arvensis	–
muricatus	–
falcatus	–

contd.....

Table 4.1 : contd....

Scientific Name	Status
Caltha	
palustris	T
Coptis	
teeta	T
Delphinium	
denudatum	T
caeruleum	–
elatum	–
brunonianum	–
Actaea	
spicata	–
Cimicifuga	
foetida	T
Paeonia	
emodi	T
Aconitum	
lucidum	–
chasmanthum	–
violaceum	–
heterophyllum	T
palmatum	T
deinorrhizum	–
balfourii	–
falconeri "var. I	–
spicatum	–
laciniatum	–
ferox	T
elwesii	–
lethal	–
napellus	–
Dillieniaceae	
Dillenia	
indica	PS
Magnoliaceae	
Michelia	
champaca	PS
nilagirica	PS
montana	S
Illicium	
griffithii	–
Annonaceae	
Ucaria	
narum	S
Artabotrys	
odoratissimus	C
Canangium	
odoratum	–
Annona	
Squamosa	C
reticulata	C
Sageraea	
laurifolia	–
Polyalthia	
longifolia	PS
simiarum	PS
thwaites	–
Menispermaceae	
Tinospora	
malabarica	T
cordifolia	T
Anamirta	
cocculus	–
Tiliacora	
acuminata	S
Cocculus	
hirsutus	S
pendulus	S
laurifolius	S
macrocarpus	S
Pachygone	
ovata	–
Pericampylus	
glaucus	–
Stephania	
hernandifolia	S
glabra	–

contd.....

Table 4.1 : contd....

Scientific Name	Status	Scientific Name	Status
Cissampelos		**Cruciferae**	
pareira	S	*Nasturtium*	
Tinospora sinensis		palustre	S
Stephania japonica		indicum	PS
Berberidaceae		montanum	–
Berberis		*Cardamine*	
petiolaris	–	impatiens	–
aristata	–	*Capsella*	
lycium	–	bursa-pastoris	–
asiatica	T	*Lepidium*	
Mahonia		sativum	S
napaulensis	T	draba	S
Podophyllum		crassifolium	S
emodi	T	latifolium	–
Nymphaeaceae		ruderale	S
Nymphaea		perfoliatum	S
alba	T	*Cochlearia*	
rubra	T	flava	–
pubescens	T	*Crambe*	
stellata	T	cordifolia	–
Euryale		*Raphanus*	
ferox	T	sativus	C
Nelumbo		*Brassica*	
nucifera	T	oleracea	C
Nymphaea nouchali		campestris" var. apa	C
		nigra	C
Papaveraceae		**Capparidaceae**	
Argemone		*Cleome*	
mexicana	PS	monophylla	–
Meconopsis		brachycarpa	–
aculeata	–	icosandra	PS
napaulensis	T	felina	–
Papaver		heptaphylla	–
somniferum	C	chelidonii	–
Fumariaceae		gynandra	–
Corydalis		*Macrua*	S
govaniana	S	arenaria	–
ramosa	–		

contd.....

Table 4.1 : contd....

Scientific Name	Status	Scientific Name	Status
Crataeva		**Flacourtiaceae**	
murvala	PS	*Flacourtia*	
Cadaba		cataphracta	S
farinosa	–	ramontchi	S
trifoliata	–	sepiaria	S
Capparis		*Camellia*	
spinosa		thea	PS
heyneana	PS	*Gordonia*	
decidua	PS	obtusa	–
grandis	S	**Dipterocarpaceae**	
sepiaria	S	*Dipterocarpus*	
zeylanica	S	turbinatus	PS
Resedaceae		tuberculatus	PS
Reseda		alatus	PS
odorata	–	pilosus	
Oligomeris		*Shorea*	
subulata	–	robusta	PS
Ochradenus		tumbuggaia	–
baccatus	–	*Hopea*	
Violaceae		odorata	S
Viola		*Vateria*	
serpens	T	indica	S
odorata	–	**Malvaceae**	
cinerea	–	*Althaea*	
diffusa	–	officinalis	–
patrini	–	rosea	–
tricolor	–	ludwigii	–
biflora	T	*Malva*	
sylvestris	–	sylvestris	S
Ionidium		rotundifolia	S
enneaspermum	S	parviflora	–
Cochlospermaceae		verticillata	–
Cochlospermum		*Malvastrum*	
gossypium	PS	coromandelianum	S
Bixaceae		*Sida*	
Bixa		veronicaefolia	S
orellana	C	rhomboidea	S
		spinosa	S

contd.....

Table 4.1 : contd....

Scientific Name	Status
acuta	S
rhombifolia	S
cordifolia	S
Abutilon	
indicum	S
hirtum	–
theophrasti	S
glaucum	–
asiaticum	–
Malachra	
capitata	S
Urena	
lobata	S
sinuata	S
repanda	S
Pavania	
odorata	
zeylanica	–
Hibiscus	
furcatus	S
micranthus	S
cannabinus	S
sabdariffa	S
abelmoschus	S
esculentus	S
tiliaceus	S
rosa-sinensis	C
lampas	S
trionum	S
surattensis	S
manihot	S
mutabilis	C
Thespesia	
populnea	PS
Gossypium	
herbaceum	S
arboreum" var.	PS
barbadense	–
hirsutum	S
Kydia	
calycina	PS
Bombacaceae	
Adonsonia	
digitata	PS
Cynocardia	
odorata	T
Hydnocarpus	
wightiana	–
venenata	–
anthelmintica	–
Taraktogenos	
kurzii	D
Pittosporaceae	
Pittosporum	
napaulense	–
Polygalaceae	
Polygala	
crotalarioides	–
chinensis	S
telephioides	S
glomerata	
sibirica	S
Frankeniaceae	
Frankenia	
pulverulenta	–
Caryophyllaceae	
Saponaria	
vaccaria	S
Polycarpea	
corymbosa	S
Vaccaria pyramidata	S
Portulacaceae	
Portulaca	
oleracea	S
quadrifida	PS
tuberosa	S

contd.....

Table 4.1 : contd....

Scientific Name	Status
Tamaricaceae	
Tamarix	
troupii	S
dioica	S
aphylla	–
gallica	S
Myricaria	
elegans	–
germanica	–
Elatinaceae	
Bergia	
odorata	–
Hyperiacaceae	
Hypericum	
patulum	T
perforatum	–
humifusum	–
sampsoni	–
japonicum	T
chinense	–
Guttiferae	
Garcinia	
mangostana	C
indica	–
morella	–
xanthochymus	S
dulcis	–
cornea	–
cowa	S
heterandra	–
Ochrocarpa	
longifolius	S
Calophyllum	
inophyllum	S
apetalum	–
elatum	–
Mesua	
ferea	PS
Ternstroemiaceae	
Schima	
wallichii	PS
monogynum	–
coca	–
Malpighiaceae	
Hiptage	
benghalensis	C
Zygophyliaceae	
Tribulus	
terrestris	–
alatus	–
Zygophyllum	
simplex	–
coccineum	–
Fugonia	
cretica	–
Geraniaceae	
Geranium	
wallichianum	–
napalense	T
robertianum	–
ocellatum	–
lucidum	S
molle	–
pratense	–
pusillum	–
rotundifolium	–
sibiricum	–
Oxalidaceae	
Oxalis	
corniculata	PS
acetosella	PS
Biophytum	
sensitivum	PS
Averrhoa	
carambola	C
bilimbi	–

contd.....

Table 4.1 : contd....

Scientific Name	Status	Scientific Name	Status
Balsaminaceae		*Clausena*	
Impatiens		excavata	T
balsamina	C	*Bombax*	
tripetala	S	ceiba	PS
chinensis	–	*Ceiba*	
Rutaceae		pentandra	PS
Evodia		**Sterculiaceae**	
Lunur-ankenda	S	*Pterygota*	
fraxinifolia	PS	alata	PS
rutaecarpa	–	*Sterculia*	
Ruta		foetida	S
graveolens	–	urens	S
tuberculata	–	balanghas	–
Peganum		rubiginosa	–
harmala	–	*Pterocymbium*	
Dictamnus		javanicum	S
albus	–	*Scaphium*	
Zanthoxylum		affine	–
alatum	–	wallichii	–
acanthopodium	S	*Helicteres*	
oxyphyllum	S	isora	PS
hamiltonianum	–	*Peterospermum*	
budrunga	S	suberifolium	S
Toddalia		acerifolium	PS
asiatica	S	heyneanum	–
Vepris		*Pentapetes*	
bilocularis	–	phoenicea	–
Skimmia		*Eriolaena*	
laureola	T	quinquelocularis	S
Glycosmis		*Melochia*	
cochinchinensis	–	corchorifolia	S
Acronychia		*Abroma*	
laurifolia	S	augusta	PS
Murraya		*Guazuma*	
koenigii	PS/C	tomentosa	PS
paniculata	C	*Theobroma*	
		cacoa	C

contd.....

Table 4.1 : contd....

Scientific Name	Status	Scientific Name	Status
Buettneria		**Erythroxylaceae**	
herbaca	–	*Erythroxylon*	
Tilinceae		pentaphylla	S
Grewia		wampi	–
tiliaefolia	S	*Limonia*	
asiatica	S	crenulata	S
sclerophylla	–	*Luvunga*	
villosa	–	scandens	S
hirsuta	S	*Paramignya*	
tenax	–	monophylla	S
umbellata	–	longispina	–
paniculata	–	*Atalantia*	
microcos	–	monophylla	T
Triumfetta		*Citrus*	
bartramia	S	medica	C
semitriloba NK	–	medica var. proper	C
Corchorus		medica var. limonum	–
capsularis	C	medica var. acida	C
olitorius	C	medica var. limetta	C
trilocularis	C	aurantium	
fascicularis	–	aurantium var. proper	–
depressus	–	aurantium var. bigaradia	–
Elaeocarpaceae		aurantium var. bergamia	
Elaeocarpus		maxima	C
ganitrus	T	*Feronia*	
oblongus	–	elephantum	S
serratus	S	*Aegle*	
tuberculatus	S	marmelos	PS
Linaceae		**Simaroubaceae**	
Linum		*Ailanthus*	
usitatissimum	C	glandulosa	S
perenne	–	excelsa	C
strictum	–	malabarica	S
Reinwardtia		*Samadera*	
trigyna	S	indica	–
Mugonia		indica var. lucida	–
mystax S	–	*Picrasma*	
		quassioides	S

contd.....

Table 4.1 : contd....

Scientific Name	Status
javanica	S
Brucea	
sumatrana	–
Eurycoma	
longifolia	–
Balanites	
aegyptiaca	T
roxburghii	–
Ochnaceae	
Ouratea	
angustifolia	–
Ochna	
pumila	S
squarrosa	–
Buseraceae	
Boswllia	
serrata	PS
glabra	–
Garuga	
pirnata	PS
Commiphora	
mukul	T
agallocha	D
stocksiana	–
Canarium	
commune	–
strictum	S
bengalense	S
Meliaceae	
Turracea	
villosa	–
Naregamia	
alata	–
Azadirachta	
indica	PS
Melia	
azedarach	PS
composita	S
Dysoxylum	
hamiltonii	S
malabaricum	S
Sandoricum	
indicum	–
Aglaia	
odoratissima	–
roxburghiana	–
Aphanamixis	
polystachya	PS
Amoora	
cucullata	–
Walsura	
piscidia	–
heynea	S
trijuga	S
Carapa	
moluccensis	T
obovata	T
Soymida	
febrifuga	S
Chukrasia	
tabularis	PS
tooba atach	PS
Chloroxylon	
swietenia	PS
Icacinaceae	
Ximenia	
americana	–
Olacaceae	
Olax	
scandens	–
nana	–
Sarcostigma	
kleinii	–
Celastraceae	
Euonymus	
tingens	–

contd.....

Table 4.1 : contd....

Scientific Name	Status
javanicus	S
Kokoona	
zeylanica	–
Celastrus	
paniculata	T
Gymnosporia	
spinosa	–
royleana	–
Elaeodendron	
glaucum	S
Salacia	
oblonga	–
reticulata	–
Rhamnaceae	
Ventilago	
maderaspatana	S
talyculata	–
Berchemia	
lineata	–
Zizyphus	
mauritiana	S
trinervia	S
nummularia	S
sativa	–
rugosa	S
oenoplia	S
Rhamnus	
dahuricus	–
wightii	–
purpureus	–
triqueter	–
nipalensis	–
Gonania	
leplostachya	S
Viitaceae	
Vitis	
quadrangularis	S
adnata	–
latifolia	–
vinifera	–
indica	–
setosa	–
carnosa	–
araneosa	–
pedata	S
tomentosa	–
repens	–
pallida	–
Leea	
macrophylla	–
crispa	S
indica	PS
robusta	–
aequata	–
Sapindaceae	
Cardiospermum	
halicacabum	S
Aesculus	
indica	S
hippocastanum	S
Allophylus	
serratus	S
Schleichera	
trijuga	PS
Sapindus	
trifoliatus	–
mukorossi	S
Litclu	
chinensis	C
Nephelium	
longana	S
lappaceum	–
Acer	
pictum	–
Dodonaea	
viscosa	PS

contd.....

Table 4.1 : contd....

Scientific Name	Status	Scientific Name	Status
Anacardiaceae		**Connaraceae**	
Rhus		*Connarus*	
parviflora	–	monocarpus	–
semialata	S	*Rourea*	
wallichii	–	santaloides	–
insignis	–	*bofalaria*	–
succedanea	–	burhia	–
Pistacia		prostrata	–
integerrima	T	albida	–
Mangifera		verrucosa	–
indica	S	juncea	–
caesia	–	medicaginea	–
Anacardium		trifoliastrum	–
occidentale	PS	retusa	–
Buchanania		sericea	–
lanzan	PS	*Trigonella*	
Melanorrhoea		occulta	–
usitata	–	foenum-graecum	S
wallichii	S	polycerata	–
curtisii	–	corniculata	–
Lamea		*Melilotus*	
coromendelica	S	indica	S
Semecarpus		officinalis	S
anacardium	PS	alba	–
Holigarna		*Cyamopsis*	
arnottiana	S	tetragonoloba	S
longifolia	–	*Indigofera*	
Spondias		ilinifolia	S
pinnata	PS	glandulosa	–
Coriariaceae		enneaphylla	S
Coriaria		aspalathoides	–
nepalensis	T	trifoliata	–
Moringaceae		oblongifolia	–
Moringa		tinctoria	–
oleifera	C	pulchella	–
coneanensis	–	trita	–
		articulata	–
		glabra	–

contd....

Table 4.1 : contd....

Scientific Name	Status
Pseudarthri	
viscid	–
Uraria	
picta	–
lagopoides	S
hamosa	–
Alysicarpus	
longifolius	S
arachis	
hypogaea	C
Ougenia	
oojeinensis	S
Desmodium	
tihaefolium	S
gangeticum	–
polycarpum	S
trifolium	–
lasiocarpum	–
retroflexum	S
pulchellum	
Abrus	
precatorius	T
Cicer	
arietinum	C
Lathyrus	
sativus	C
aphaca	–
pratensis	
Pisum	
sativum	C
Glycine	
soja	–
Teranmus	
labialis	–
Mucuna	
monosperma	–
gigantea	–
prurita	T
Erythrina	
indica	PS
Psoralea	
corylifolia	–
Colutea	
nepalensis	–
Mundulea	
suberosa	–
Tephrosia	
parpurea	S
villosa	S
petrosa	–
Glycyrrhiza	
glabra	S
Millettia	
auriculata	S
pachycarpa	S
Adinobotrys	
atropurpureus	–
Sesbania	
aegyptiaca	C
aculeata	C
grandiflora	C
Astragalus	
tribuloides	–
hamosus	S
multiceps	–
strobiliferous	–
Taverniera	
cuneifolia	–
Lens	
esculenta	C
Athagi	
camelorum	–
Zornia	
diphylla	S
Smithia	
conferta	S

contd.....

Table 4.1 : contd....

Scientific Name	Status
sensitiva	–
Ornocarpum	
sennoides	–
lunatus	–
aconitifolius	–
adenanthus	–
catiang	–
Clitoria	
ternatea	C
Dolichos	
biflorus	–
lablab	C
Rhynehosia	
minima	–
Cajanus	
cajans	C
Atylosia	
searabaeoides	S
Cylista	
seariosa	–
Flemingia	
strobilifera	–
chappar	S
grahamiana	–
congesta	S
tuberosa	–
nana	–
Dalbergia	
sisso	PS
multiflora	–
lanceolaria	S
volubilis	S
spinosa	S
famarindifolia	S
laifolia	T
Pterocarpus	
santalinus	T
indicus	D
marsupium	T
Pongamia	
pinnata	PS
Derris	
scandens	S
elleptica	PS
uliginosa	
Sophora	
tomentosa	–
mollis	S
griffithii	–
"var. roxburghi	–
suberosa	–
Butea	
monosperma	PS
superba	S
Canavalia	
virosa	–
ensiformis	–
Pueraria	
tuberosa	–
Phaseolus	
trilobus	–
radiatus	C
mungo	C
Caesalpiniaceae	
Caesalpinia	
crista	S
jayabo	–
nuga	–
sapan	–
pulcherrima	C
sepiaria	–
digyna	S
coriaria	–
Delonix	
elata	C
Wagatea	
spicata	–

contd.....

Table 4.1 : contd....

Scientific Name	Status
Cassia	
fistula	PS
occidentalis	PS
sophera	PS
obtusifolia	–
auriculata	S
obovata	–
alata	PS
glauca	–
absus	–
mimosoides	–
pumila	–
angustifolia	T/C
javanica	–
tora	PS
Cynometra	
mimosoides	–
cauliflora	–
Hardwickia	
pinnata	PS
Saraca	
indica	C
Ceratonia	
siliqua	–
Haematoxylon	
campechianum	–
Tamarindus	
indica	PS
Humboldtia	
vahliana	S
Bauhinia	
tomentosa	–
racemosa	PS
retusa	PS
vahlii	PS
purpurea	PS
variegata	PS
malabarica	–
macrostachya	–
Mimosaceae	
Neptunia	
oleracea	–
Xylia	
dolabriformis	S
Entada	
scandens	S
Adenanthera	
pavonia	T
Prosopis	
spicigera	PS
Dichrostachys	
cinerea	–
Leuceena	
glauca	PS
Mimosa	
pudica	PS
runicaulis	PS
Acacia	
farnesiana	S
nilotica	PS
leucophloea	
catechu	PS
ferruginea	–
senegal	–
modesta	–
rugata	–
caesia	–
pennata	S
suma	S
Albizzia	
lebbeck	PS
odoratissima	PS
julibrissin	–
amara	–
procera	PS
stipulata	PS
Pithecellobium	
bigeminum	PS
dulce	PS

contd.....

Table 4.1 : contd....

Scientific Name	Status	Scientific Name	Status
Rosaceae		gallica	–
Prunus		alba	C/S
amygdalus	S	indica	–
persica	S	banksiae	–
armeniaca	–	multiflora	–
cerasus	–	*Cydonia*	
avium	C/S	vulgaris	–
cerasoides	S	*Pyrus*	
communis	–	malus	C
domestica	–	communis	–
insititia	–	*Eriobotrya*	
cornuta	–	japonica	C
mahaleb	–	*Cotoneaster*	
undulata	–	nummularia	S/r
triflora	–	bacillaris	–
Prinsepia		microphylla	S
utilis S	S	var. buxifolia	–
Rubus		**Saxifragaceae**	
moluccanus	–	*Bergenia*	
saxatilis	–	ligulata	–
fruticosus		*Dichora*	
Geum		febrifuga	S
urbanum	–	*Ribes*	
alatum	–	orientale	S
Potentilla		nigrum	–
nepalensis	S	*Kalanchoe*	
supina	–	pinnata	–
anserina	S	spathulata	–
fruticosa	S	laciniata	–
reptans	–	*Sedum*	
fragarioides	–	asiaticum	S
kleiniana	S	multicaule	–
sericea	S	**Droseraceae**	
Agrimonia		*Drosera*	
eupatoria	S	lunata	–
Rosa		indica	–
damascena	C	burmanni	T
centifolia	–		

contd.....

Table 4.1 : contd....

Scientific Name	Status
Hamamelidaceae	
Exbacklandia populnea	S/C
Rhizophoraceae	
Rhizophora	
mucronata	PS
Ceriops	
candolleana	PS
Kandelia	
candel	T
Combretaceae	
Terminalia	
catappa	S
belerica	PS
chebula	S
citrina	S
arjuna	PS
tomentosa	PS
paniculata	–
pyrifolia	–
bialata	–
oliveri	–
myriocarpa	PS
coriacea	
pallida	D
Calycopteris	
floribunda	D
Anogeissus	
latifolia	PS
Quisqualis	
indica	C
Lecythedaceae	
Barringtonia	
racemosa	–
acutangula	PS
speciosa	–
Careya	
arborea	PS
Myrtaceae	
Myrtus	
communis	–
Melaleuca	
leucadendron	C
Luffa	
aegyptiaca	C
acutangula	C
var. amara	–
echinata	–
Benincasa	
hispida	C
cerifera	–
Momordica	
charantia balsamina	–
dioica	C
cochinchinensis	C
tuberosa	–
Cucum	
trigonus	–
melo	C/PS
"var. momordica	–
"var. utilissimus	–
prophetarum	–
sativus	–
Citrullus	
colocynthis	–
vulgaris	C
"var fistulosus	–
Coccinia	
indica	S
Cucurbita	
maxima	C
pepo	C
Bryonopsis	
laciniosa	C
Melothria	
maderaspatana	S
perpusilla	–

contd.....

Table 4.1 : contd....

Scientific Name	Status	Scientific Name	Status
heterophylla	S	Bupleurum	–
Blastania		*Eucalyptus*	
garcini	–	citriodora	PS
Kedostris		globulus	–
rostrata	–	*Psidium*	
Corallocarpus		guyava	C
epigaeus	–	*Jambosa*	
Zanonia		jambos	PS
indica	–	*Syzygium*	
Datiscaceae		operculatum	S
Datisca		cumini	PS
cannabina	–	spicatum	–
Cactaceae		hemispherican	–
Opuntia		*Pimenta*	
coccinellifera	–	acris	–
monacantha	–	**Melastomaceae**	
stricta	–	*Mdemecylon*	
nigricans	–	umbellatum	–
dillenii	S	amplexicaule	S
Ficoidaceae		angustifolium	–
Trianthema		*Melastoma*	
portulacastrum	S	malabathricum	S
pentandra	–	decemfidum	–
decandra	–	*Osbeckia*	
Mollugo		nepalensis	S
hirta	S	crinita	S
oppositifolia	–	*Ammannia*	
pentaphylla	–	baccifera	
cerviana	PS	auriculata	S
nudicaulis	–	**Lythraceae**	
Giesekia		*Woodfordia*	
pharmacoides	–	fruticosa	S
Umbelliferae		*Lawsonia*	
Hydrocotyle		inermis	PS
asiatica	PS	*Lagerstroemia*	
javanica	PS	speciosa	–
Eryngium		indica	c
caeruleum	–	*Sonneratia*	
		caseolaris	D

contd.....

Table 4.1 : contd....

Scientific Name	Status
Punicaceae	
Punica	
granatum	C
Onagraceae	
jussieua	
suffruticosa	S
repens	–
Trapa	
bispinosa	S
Samydaceae	
Casearia	
graveolens	S
esculenta	S
tomentosa	–
Cariaceae	
Carica	
papaya	C
Passifloraceae	
Passiflora	
foetida	C
edulis	–
Adenia	
palmata	C
Cucurbitaceae	
Trichosanthes	
palmata	C
cordata	C
dioica	C
nervifolia	C
cucumerina	–
anguina	C
wallichiana	–
Gymnopetalum	
cochinchinense	–
Laganaria	
vulgaris	C
falcatum	–
jucundum	–
Apium	
graveolens	–
Carum	
carvi	C
bulbocastanum	–
roxburghianum	–
copticum	C
Pimpinella	
heyneana	C
sexifraga	–
diversifolia	–
stocksii	–
Seseli	
indicum	S
Pycnocycla	
aucheriana	–
Foeniculum	
capillaceum	C
Prangos	
pabularia	–
Angelica	
glauca	–
Ferula	
narthex	C
jaeschkeana	–
toctida	–
Peucedamum	
graveolens	–
grande	–
dhana	–
aucheri	–
nagpurense	
Heracleum	
wallichii	–
Zosimia	
orientialis	–

contd.....

Table 4.1 : contd....

Scientific Name	Status
Coriandrum	
sativum	C
Nothopanax	
fruticosum	–
Hedera	
helix	C
Cuminum	
cyminum	–
Daucus	
carota	C
Psammogeton	
biternatum	–
Alangiaceae	
Alangium	
lamarckii	S
Caprifoliaceae	
Sambucus	
eburus	–
javanica	S
Viburnum	
foetidum	S
Lonicera	
glauca	–
Rubiaceae	
Sarcocephalus	
missionis	–
cordatus	–
Anthocephalus	
chinensis	PS
Adina	
cordifolia	PS
Nauclea	
sessilifolia	PS
Mytragyna	
parvifolia	
Uncaria	
gambir	–
Hymenodictyon	
excelsum	PS
Paederia	
foetida	T
Pavetta	
indica	PS
Randia	
dumetorum	S
uliginosa	S
Rubia cordifolia	
Vangueria	S
spinosa	S
Morinda	
citrifolia	PS
Ophiopogon	
serpentinum	–
Cuises	
officinalis	–
Oldenlandia	
corymbosa	S
Psychotria	
ipecacuana	S
Ophiorrhiza	
mungo's	S
Ixora	
parviflora	C
cocinea	C
Mussaenda	
frondosa	PS
Valarianaceae	
Nardostachys	
jatamansi	T
Valeriana	
hardwickii	–
officinalis	–
Asclepiacaceae	
Hemidesmus	
indicus	T

contd.....

Table 4.1 : contd....

Scientific Name	Status
Cryptolepis	
buchanani	T
Cryptostegia	
grandiflora	S
Periploca	
aphylla	S
Secamone	
emetica	S
Glossonema	
varians	S
Oxystelma	
esculentum	S
Calotropis	
gigantea	PS
procera	PS
Asclepias	
curassavica	PS
Pentatropis	
cynanchoides	–
microphylla	–
Pergularia	
extensa	–
Daemia	
cordata	–
Holostemma	
annulare	–
Cynanchum	
arnottianum	–
Sarcostemma	
brevistigma	–
brunonianum	–
intermedium	–
stocksii	–
Gymbema	
sylvestre	T
Marsdenia	
roylei	–
volubilis	–
Leptadenta	
reticulata	–
Tylophora	
fasciculata	–
asthmatica	–
tenuis	–
Cosmostigma	
racemosum	–
Dregea	
valubilis	S
Caropegia	
bulbosa	S
tuberosa	–
Caralluma	
edulis	–
Boucerosia	
aucheriana	–
Loganiaceae	
Fagraea	
racemosa	S
Strychnos	
colubrina	S
nux-vomica	T
potatorum	T
bourdilloni	–
cinnamomifolia	–
Cyrtophyllum	
peregrinum	–
Gentianaceae	
Exacum	
tetragonum	S
bicolor	–
pedunculatum	S
lawii	–
Enicostemma	
littorale	–
Erythraea	
roxburghii	–

contd.....

Table 4.1 : contd....

Scientific Name	Status
Happea	
dichotoma	–
Canscora	
diffusa	S
decussata	S
Gentiana	
tenella	–
kurroo	–
decumbens	S
dahurica	–
Swertia	
purpurascens	–
paniculata	–
chirata	T
angustifolia	T
angustifolia Vr	–
decussata	–
alata	–
lawii	–
Limnanthemum	
nymphaeoides	PS
Menyanths	
trifoliata	–
Hydrophyllaceae	
Hydrolea	
zeylanica	–
Boraginaceae	
Cordia	
obliqua	PS
wallichii	–
rothii	PS
vestita	–
macleodii	–
Ehretia	
aspera	–
microphylla	–
Coldenia	
procumbens	–
Rotula	
acquatica	–
Heliotropium	
eichwaldi	–
tuberculosum	
strigosum	PS
brevifolium	–
indicum	–
Trichodesma	
indicum	–
africanum	PS
zeylanicum	–
Cynoglossum	
glochidiatum	PS
Macrotomia	
benthami	–
perennis	–
Onosma	
echioides	–
bracteatum	
Caccinia	
glauca	–
Lithospermum	
officinale	–
arvense	–
Convolvulaceae	
Argyreia	
speciosa	S
fulgens	–
Lettsomia	
aggregata	S
Calonyction	
bona-nox	–
muricatum	–
Quamoclit	
pinnata	PS
coccinea	–
vulgaris	–

contd....

Table 4.1 : contd....

Scientific Name	Status	Scientific Name	Status
Erycibe		**Solanaceae**	
paniculata	–	*Solanum*	
Rivea		nigrum	PS
ornata	PS	dulcamara	T
Ipomoea		spirale	–
hederacea	S	verbascilolium	PS
uniflora	–	ferox	S
paniaelata	–	indicum	S
batatas	C	melongena	C
pes-tigridis	PS	xanthocarpum	S
reniformis	–	trilobatum	–
obscura	–	gracilipes	–
sepiaria	–	torvum	S
reptans	C	incanum	–
campanulata	–	sarattense	S
pes-caprae	PS	*Physalis*	
dissecta	–	minima	S
tuberosa	–	minima var. indica	PS
dasysperma	–	angulata	–
hispida	–	peruviana	S
Operculina		capsicum	–
turpethum	–	frutescens	–
Merremia		annuum	PS
vitifolia	PS	minimum	–
tridentata	–	*Withania*	
Convolvulus		somnifera	T
arvensis	PS	coagulans	–
glomeratus	–	*Nicandra*	
spinosus	–	physaloides	–
Evolvulus		*Lycium*	
alsinoides	PS	barbarum	–
Cressa		ruthenicum	–
cretica	–	*Atropa*	
Cuscuta		belladonna	T
reflexa	PS	*Datura*	
hyalina	–	stramonium	S
chinensis	–	fastuosa var al.	C/S
		metal	S

contd.....

Table 4.1 : contd....

Scientific Name	Status	Scientific Name	Status
Scopolia		roxburghii	–
lurida	–	*Moniera*	
Physoclaina		cuneifolia	–
praealta	–	*Artanema*	
Hyoscyamus		sesamoides	–
niger	T	*Curanga*	
muticus	–	amara	S
reticulatus	–	*Veronica*	
Nicotiana		anagallis	S
tabacum	C	beccabunga	–
rustica	–	*Striga*	
Scrophulariaceae		lutea	–
Torenia		orobanchoides	
asiatica	PS	*Sopubia*	
Vandellia		delphinifolia	–
pysidaria	PS	*Pedicularis*	
pedunculata	–	pectinata	–
Bonnaya		siphonantha	–
reptans	PS	**Orobanchaceae**	
Scoparia		*Cistanche*	
dulcis	PS	tubulosa	–
Picrorrhiza		*Orobanche*	
kurrooa	T	aegyptiaca	T
Verbascum		**Lentibulariaceae**	
thapsus	S	*Utricularia*	
Celsia		bifida	s
coromandeliana	S	**Bignoniaceae**	
Linaria		*Oroxylum*	
ramosissima	PS	indicum	PS
Schweinfurthia		*Tecomella*	
sphaerocarpa	–	undulata	C
Lindenbergia		*Dolichandrone*	
urticaefolia	PS	spathacea	S
Stemodia		falcata	–
viscosa	–	*Heterophragma*	
Limnophila		roxburghii	
gratissima	–	*Stereospemum*	
gratioloides	–	Chelonoides	PS

ccntd.....

Table 4.1 : contd....

Scientific Name	Status	Scientific Name	Status
suaveolens	PS	*Haplanthus*	
Radermachera		verticillaris	–
xylocarpa	–	tentaculatus	–
Amphicome		*Gymnostachyum*	
emodi	NK	febrifugum	–
Tecoma		*Phlogacanthus*	
stans	C	thyrsiflorus	PS
Crescentia		*Crossandra*	
cujete	C	undulaefolia	–
Pedaliaceae		*Asystasia*	
Martynia		gangetica	–
annua	S	*Lepidagathis*	
Pedalium		cristata	S
murex	–	trinervis	S
sesamum	C	hamiltoniana	–
indicum	C	*Justica*	
Acanthaceae		gendarussa	PS
Auriculatus		procumbens	–
ciliatus	–	*Adhatoda*	
Blepharis		vasica	PS/C
edulis	S	*Rhinacanthus*	
sindica		communis	–
Acanthus		nasuta	–
ilicifolius	PS	*Ecbolium*	
Barleria		linneanum	–
prionitis	T	*Graptophyllum*	
noctiflora	–	pictum	–
cristata	T	*Rungia*	
cristata var. dichotoma	–	repens	–
strigosa	T	parviflora	S
courtallica	–	*Dicliptera*	
longiflora	–	roxburghiana	S
Neuracanthus		*Peristrophe*	
sphaerostachys	–	bicalyculata	–
Andrographis		*Cardanthera*	
paniculata	T	uliginosa	S
echiodes	–	*Asteracantha*	
		longifolia	S

contd.....

Table 4.1 : contd....

Scientific Name	Status	Scientific Name	Status
Ruellia		agnus-castus	–
prostrata	S	pubescens	–
suffruitcosa	–	leucoxylon	–
Daedalacanthus		*Clerodendrum*	
roseus	S	inerme	S
Strobilanthes		phlomidis	C/S
callosus	S	serratum	S
Verbenaceae		speciosium	PS
Lantana		siphonanthus	S
indica	PS	*Avicennia*	
aculeata	PS	officinalis	PS
Lippia		tomentosa	–
nodiflora	PS	**Labiatae**	
Verbena		*Ocimum*	
officinalis	C	canum	C
Callicarpa		basilicum	C
arborea	S	gratissimum	–
lanata	–	sanctum	c
macrophylla	S	*Geniosporum*	
cana	–	prostratum	–
Stachytarpheta		*Orthosiphon*	
indica	S	stamineus	S
Tectona		*Coleus*	
grandis	PS	amboinicus	C
Premna		*Anisochilus*	
integrifolia	S	carnosus	–
tomentosa	–	*Lavandula*	
latifolia	PS	bipinnata	–
esculenta	–	*Pogostemon*	
herbacea	S	plectranthoides	S
Gmelina		purpurascens	S
arborea	PS/C	parviflorus	–
asiatica	–	*Colebrookia*	
Vitex		oppositifolia	S
trifolia	–	*Mentha*	
negundo	PS	viridis	–
peduncularis	PS	piperita	C
glabrata	–	sylvestris	C

contd. ...

Table 4.1 : contd....

Scientific Name	Status	Scientific Name	Status
arvensis	C	*Brunella*	
Lycopus		vulgaris	S
europaeus	–	*Marrubium*	
Origanum		vulgare	–
majorana	–	*Anisomeles*	
vulgare	–	indica	S
Thymus		malabarica	–
serpyllum	–	*Stachys*	
Hyssopus		parviflora	–
officinalis	–	*Leonurus*	
Micromeria		sibiricus	S
capitellata	S	*Roylea*	
Calamintha		elegans	–
clinopodium	S	*Otostegia*	
Melissa		limbata	–
parviflora	S	aucheri	–
Perowskia		*Leucas*	
abrotanoides	–	cephalotes	–
atriplicifolia	–	zeylanica	–
Meriandra		aspera	PS
strobilifera	–	linifolia	–
bengalensis	–	urticeafolia	–
Salvia		stelligera	–
moorcroftiana	–	*Leonotis*	
lanata	C	nepetaefolia	–
plebeia	–	*Eremostays*	
aegyptiaca	–	vicaryi	–
cabulica	–	acanthocalyx	–
spinosa	–	*Hyptis*	
officinalis	–	suaveolens	S
Nepeta		*Ajuga*	
elliptica	S	bracteosa	
ciliaris	–	*Zataria*	
ruderalis	–	multiflora	–
glomerulosa	–	*Ziziphora*	
Dracocephalum		clinopodioides	–
moldavicum	–	tenuior	–
Lallemantia		*Hymenocrater*	
royleana	–	sessilifolius	–

contd.....

Table 4.1 : contd....

Scientific Name	Status
Tenucrium	
stocksianum	–
scordium	–
Plantaginaceae	
Plantago	
major	C
lanceolata	–
amplexicaulis	–
ovata	C
psyllium	–
ciliata	–
lagocephala	–
Nyctaginaceae	
Boerhavia	
diffusa	PS
repens	PS
Pisonia	
aculeata	–
morindaefolia	–
Mirabilis	
jalapa	S
Amaranthaceae	
Celosia	
argentea	S
"var. cristata	
Digere	
arvensis	S
Amaranthus	
spinosus	PS
paniculatus	–
gangeticus	PS
"var. tristis	–
virdis	–
blitum	–
"var. oleracea	
Aerva	
tomentosa	PS
lanata	PS
Acyranthes	
aspera	PS
bidentata	PS
Alternanathera	
sessilis	PS
Chenopodiaceae	
Chenopodium	
album	PS
botrys	–
ambrosiodes	PS
Beta	
vulgaris	C
Spinacia	
oleracea	C
Kochia	
indica	–
sedoides	–
scoparia	–
Arthrocnemum	
indicum	–
Salicornia	
brachiata	–
Suaeda	
fruticosa	PS
monoica	–
Salsola	
kali	–
foetida	–
Basella	
rubra	C
Haloxylon	
salicornicum	–
recurvum	–
Phytolaccacesae	
Phytolacca	
acinosa	S

contd.....

Table 4.1 : contd....

Scientific Name	Status	Scientific Name	Status
Polygonaceae		scutatus	–
Calligonum		Rumex maritimus L.	PS
polygonoides	S	Rumex vesicarius L.	–
Pteropyrum		**Aristolochiaceae**	
olivierii		*Bragantia*	
Polygonum		wallichii	–
aviculare	–	tomentosa	–
plebejum	S	*Aristolochia*	
viviparum	–	bracteata	–
glabrum	–	indica	T
persicaria	–	tagala	T
barbatum	PS	**Piperaceae**	
hydropiper	S	*Piper*	
punctatum	–	longum	C
molle	S	cluba	–
chinense	–	sylvaticum	
ientale	–	betle	PS
virginianum	–	nigrum	C
sphaerostachyum	–	attenuatum	–
serrulatum	–	sarmentosum	–
Fagopyrum		aurantiacum	–
cymosum	–	**Chloranthaceae**	
esculentum	PS	*Chloranthus*	
talaricum	–	officinalis	–
Rheum		brachystachys	
spiciforme	–	**Myristicaceae**	
emodi	T	*Myristica*	
webbianum	–	malabarica	–
nobile	T	fragrans	PS
Oxyria		**Lauraceae**	
digyna	–	*Cinnamomum*	
Rumex		tamala	C
maritimus	–	obtusifolium	C/S
dentatus	PS	iners	–
nepalensis	PS	zeylancum	C
vesicarius	PS	macrocarpum	–
acetosella	–	glanduliferum	S
acetosa	–		

contd.....

Table 4.1 : contd....

Scientific Name	Status	Scientific Name	Status
parthenoxylon	–	**Santalaceae**	
camphora	C	Santalum	
cassia	–	album	C
pauciflorum	–	*Osyris*	
javanicum	–	arborea	–
Machilus		**Euphorbiaceae**	
macrantha	S	*Euphorbia*	
Actinodaphne		hypericifolia	–
hookeri	S	hirta	S
Litsea		thymifolia	–
chinensis	S	microphylla	–
lyantha	S	tirucalli	C
stocksii		neriifolia	C
Lindera		nivulia	C
neesiana	S	antiquorum	C
Cassytha		royleana	C/PS
filiformis	PS	thomsoniana	–
Litsea glutinosa Lot. Pers.	PS	jhelioscopia	–
Litsea monopetala (Rc)	PS	dracunculoides	–
Elaeagnaceae		longifolia	–
Elaeagnus		granulata	–
hortensis	–	sanguinea	–
umbellata	S	turcomanica	–
latifolia	–	*Buxus*	
Hippophae		sempervirens	T
rhamnoides	S	*Bridelia*	
salicifolia	–	retusa	S
Loranthaceae		montana	S
Dendrophthoe falcata		*Cleistanthus*	
cochinchinensis	PS	collinus	S
Loranthus		*Andrachne*	
elasticus	–	cordifolia	–
falcatus	PS	*Phyllanthus*	
Viscum		reticulatus	S
album	S	maderaspatensis	S
monoicum	S	urinaria	S
orientale	–	simplex	S
articulatum	–	niruri	S

contd.....

Table 4.1 : contd....

Scientific Name	Status
Cicca	
distichа	S
Glochidion	
hohenackeri	–
zeylanicum	S
Flueggea	
virosa	S
leucopyrus	–
Sauropus	
quadrangularis	
Breynia	
rhamnoides	S
patens	S
Putranjiva	
roxburghii	PS/C
Antidesma	
bunius	T
zeylanicum	–
Jatropha	
glandulifera	–
nana	–
multifida	PS
curcas	S
gossypifolia	S
Aleurites	
moluccana	C
Bischofia	
javanica	S
Aporosa	
lindleyana	–
Croton	
reticulatus	S
oblongifolius	PS
caudatus	–
tiglium	S
Chrozophora	
rottleri	PS
prostrata	–
Acalypha	
fruticosa	–
indica	PS
hispida	–
paniculata	–
Trewia	
nudiflora	PS
Mallotus	
philippinensis	PS/S
Macaranga	
peltata	S
indica	S
Laportea	
crenulata	PS
Boehmeria	
nivea	PS
Platanaceae	
Platanus	
orientalis	S
Juglandaceae	
Juglans	
regia	C/D
Myricaceae	
Myrica	
Myrica esculenta	D
Casuarinaceae	
Casuarina	
equisetifolia	C
Cupuliferae	
Betula	
utilis	S
alnoides	–
Quercus	
incana	PS
lamellosa	S
pachyphylla	

contd.....

Table 4.1 : contd....

Scientific Name	Status	Scientific Name	Status
Corylus		kumaonensis	–
colurna	–	soongarica	–
Quercus infectoria Oliver		*Crocus*	
Salicaceae		sativus	C
Salix		*Belamecanda*	
tetrasperma	S	chinensis	C
acmophylla	–	**Amaryllidaceae**	
caprea	–	*Agave*	
alba	–	americana	C
babylonica	C	angustifolia	–
Populus		vera-cruz	–
nigra	–	*Homonoia*	
ciciata	S	riparia	S
Ceratophyllaceae		*Ricinus*	
Ceratophyllum		communis	C
demersum	S	*Baliospermum*	
Elettaria		montanum	T
cardamomum	C	axillare	T
Alpinia		*Tragia*	
galanga	T	involucrata	S
alihugas	S	*Sapium*	
calcarata	–	indicum	–
malaccensis	S	insigne	S
speciosa	–	sebiferum	S
Maranta		*Excoecaria*	
arundinaceae	C	agallocha	PS
Canna		acerifolia	–
indica	C	*Sebastiania*	
Musa		chamelaea	–
sapientum	C	*Huru*	
textilis	C	crepitans	–
Haemodoraceae		*Manihot*	
Sansevieria		utilissima	C
roxburghiana	C	*Hippomane*	
Irid		mancinella–	
Iris		**Urticaceae**	
ensata	–	*Hunulus*	
nepalensis	C	lupulus	–

contd.....

Table 4.1 : contd....

Scientific Name	Status	Scientific Name	Status
Cannabis		*Pouzolzia*	
sativa	C	indica	PS
Streblus		*Antiaris*	
asper	S	toxicaria	–
Morus		*Artocarpus*	
indica	C	hirsuta	–
alba	C	integrifolia	C
nigra	–	lakoocha	C
Ficus		*Urtica*	
gibbosa	S	parviflora	–
bengalensis	S	dioica	S
benjamina	S	pilulifera	–
retusa	S	*Curculigo*	
rumphii	S	orchioides	–
religiosa	PS	*Crinum*	
lacor	–	asiafticum	C
heterophylla	PS	latifolium	–
asperrima	–	defixum	–
hispida	PS	*Polianthes*	
cunia	S	tuberosa	C
ribes	–	**Taccaceae**	
palmata	PS	*Tacca*	
glomerata	–	pinnatifida	S
carica	–	aspera	–
arnottiana	–	**Bromeliaceae**	
dalhousiae	–	*Ananas*	
talboti	–	sativus	C
tsiela	–	**Dioscoreaceae**	
Holoptelea		*Dioscorea*	
integrifolia	PS	pentaphylla	T
Celtis		oppositifolia	T
australis	S	bulbifera	T
cinnamomea	S	triphylla	S
Terma		alata	T
orientalis	PS	var. globosa	–
Gironniera		sativa	S
reticulata	–		
Girardinia			
zeylanica	–		

contd.....

Table 4.1 : contd....

Scientific Name	Status
Liliaceae	
Smilax	
glabra	–
lanceaefolia	–
zeylanica	T
prolifera	T
merophylla	T
Asparagus	
filicinus	–
racemosus	T
adscendens	–
gonoclados	–
officinalis	
Yucca	
gloriosa	–
aloifolia	–
Aloe	
vera	T
Polygonatum	
multiflorum	–
Asphodelus	S
tenuifolius	–
Chlorophytum	
arundinaceum	S
Allium	
ascalonicum	–
cepa	C
sativum	C
schaenoprasum	–
tuberosum	–
ampeloprasum	–
Urginea	S
indica	–
coromandeliana	–
Scilla	
indica	C
Lilium	
giganteum	C
wallichianum	–
Fritillaria	
imperialis	–
roylei	–
cirrhosa	–
Colchicum	
luteum	C
Gloriosa	
superba	T
Polyanthes	
tuberosa	C
Pontederiaceae	
Monochoria	
vaginalis	S
Xyridaceae	
Xyris	
indica	–
anceps	–
pauriflora	–
Commelinaceae	
Commelina	
obliqua	S
suffruticosa	S
nudiflora	S
benghalensis	PS
salicifolia	–
Aneilema	
scapiflorum	–
Cyanotis	
tuberosa	–
axillaris	–
Floscopa	
scandens	S
Flagellariaceae	
Flagellaria	
indica	S
Juncaceae	
Luzula	
campestris	S

contd.....

Table 4.1 : contd....

Scientific Name	Status
Palmae	
Areca	
catechu	PS
nagensis	–
Loxococcus	
rupicola	–
Pinanga	
dicksonii	S
lrachis	–
Arenga	
saccharifera	–
obtusifolia	PS
Nallichia	
disticha	–
Caryota	
urens	T
mitis	–
Phoenix	
dectylifera	PS
sylvestris	PS
pusilla	–
Nannorphos	
ritchieana	–
Copernicia	
cerifera	–
Corypha	
umbraculifera	–
Borassus	
flabellifer	
Lodoicea	
seychellarum	–
Elaeis	
guineensis	–
Cocos	
nucifera	PS
schizophylla	–
yatai	–
Calamus	
rotang	T
Travancoricus	
rheedii	–
ziminalis	–
Nipa	
fruticans	T
Pandanaceae	
Pandanus	
tectorius	S
tascinalatus	
Typhaceae	
Typha	
angustata	S
elephantina	PS
laxmanni	–
Araceae	
Cryptocoryne	
spiralis	PS
Pistia	
stratiotes	PS
Lagenandra	
ovata	PS
Arisaema	
speciosum	–
tortuosum	–
leschenaultii	–
Sauromatum	
guttatum	–
Typhonium	
trilobatum	PS
Amorphophallus	
campanulatus	C
prainii	–
Synantheras	
sylvatica	–
Plesmonium	
margaritiferum	C
Remusatia	
vivipara	C

contd.....

Table 4.1 : contd....

Scientific Name	Status
Colocasia	
esculenta	C
antiquorum	c
Alocasia	
indica	C
macrorrhiza	PS
montana	–
denudata	–
Homalomena	
aromatica	–
rubescens	–
indapsus	–
officinalis	–
Rhaphidophora	
pertusa	PS
Lasia	
heterophylla	S
Pothos	
scandens	PS
cathecarti	PS
Acorus	
calamus	S
gramineus	D
Alismaceae	
Sagittaria	
sagittifolia	PS
Hydrocharitaceae	
Vallisneria	
spiralis	PS
Orchidaceae	
Desmotrichum	
fimbriatum	–
Dendrobium	
ovatum	T
Eulophia	
campestris	–
nuda	–
Cymbidium	
aloifolium	T
Vanda	
roxburghii	T
spathulata	–
tessellata	T
Saccolabium	
papillosum	T
Acampe	
wightiana	T
Zeuxine	
strateumatica	T
Orchis	
latifolia	–
Habenaria	
commelinifolia	T
Curcuma	
angustifolia	S
aromatica	C
zedoaria	S
caesia	–
amada	C
longa	C
Kaempferia	
galanga	T
angustifolia	S/T
rotunda	–
Gastrochilus	
pandurata	–
Hedychium	
spicatum	S
Amomum	
xanthioides	–
subulatum	C
aromaticum	S
costatum	–
Zingiber	
officinale	C

contd.....

Table 4.1 : contd....

Scientific Name	Status
zerumbet	S
cassumunar	S
Costus	
Speciosus	
Plumbaginaceae	
Plumbago	
zeylanica	S
rosea	S
indica	–
Sapotaceae	
Acrus	
sapota	S
Madhuca	
indica	–
longifolia	S/PS
Mimusops	
elengi	C
hixandra	S
Mamikara	
kanki	–
Sarcos stoma	
brevistina	–
Erenaceae	
Diospyros	
embryopleis	S
Symplocaceae	
Symplocos	
racemosa	S
Styraceae	
Styrax	
benzoin	S
Oleaceae	
Jasminum	
arborescens	–
grandiflorum	C/S
sambac	C
multiflorum	C
(su. pubescens)	C
humilis	–
Nyctanthes	
arbor-tristis	C/S
Schrebera	–
swietenoides	–
Salvadoraceae	
Azima	
tetracantha	–
Salvadora	
persica	S
Asteraceae	
Sonchus	
arvensis	S
brachyotes	S
Pluchia	
lanceolata	S
Vernonia	
cinera	PS
anthelmentica	PS
Elephantopus	
scaber	PS
Gnangea	
maderaspetara	S
Eupatorium	
ayapara	C
triplinerve	S
Blumea	
lacera	S
Anacyelus	
pyrethrum	S
Artimesia	
vulgaris	S
Carthamus	
tinctorius	C
Crysanthemum	
coronarium	C

contd.....

Table 4.1 : contd....

Scientific Name	Status
Eclipta	
alba	PS
prostrata	
Enhydra	
fluctuans	PS
Saussurea	
lappa	T
Guisotia	
abisinica	PS
Xanthium	
strumarium	PS
Wedelia	
calendulacea	PS
Sphaeranthes sp.	PS
Agocynaceae	
Carissa	
carandus sp.	S
Aganosma	
caryophyllata	C
dichotoma	C
Alstonia.	
scholaris	PS
Ichnocarpus	
fruitescens	PS
Holarrhene	
antidysentetica	PS
Rauwolfia	
serpentina	–
Nerium	
odorum	C
indicum	C
Wrightia	
tomentosa	PS
tinctoria	S
Thevetia	
neriifolia	C
Vallaris	
haynei	PS
solanea	PS
Plumeria	
acutifolia	C
rubra	C
Ervatania	
coronaria	C
Polypodiacesae	
Cinbotium	
barometz	–
Stenoloma	
chinensis	–
Adiantum	
lunulatum	T
eaudatum	T
eapillus-veneris	T
aethiopicum	–
venustum	–
pedatum	–
flabellulatum	–
Cheilanthes	
tenuifolia	S/D
Pteris	
aquilina	T
Asplenium	
adiantum-nigrum	T
ruta-muraria	S/P
trichomanes	–
falcatum	–
Athyrrium	
filix-foemina	T
Actiniopteris	
dichotoma	–
Aspidium	
polymorphum	T
Drynaria	
quercifolia	PS
Pleopeltis	
lanceolata	–

contd.....

Table 4.1 : contd....

Scientific Name	Status
Lygodium	
flexuosum	PS
japonicum	–
Osmundaceae	
Osmunda	
regalis	D
Ophioglossaceae	
Ophioglossum	
vulgatum	T
Helminthostachys	
zeylanica	T
Cotrychium	
lunaria	–
ternatum	–
Equisetaceae	
Equisetum	
debile	T
Polypodiaceae	
1. *Actiniopteris australis* (l.f.) L.	–
2. *Adiantum capillus* veneris L.	–
3. *Adiantum incisum* Forsk.	–
syn	
Adiantum philippense	–
Adiantum venustum	–
Polypodiaceae	
Drynaria quercifolia	S
Marsileaceae	
Marsilea minata	PS
Salviniaceae	
Azolla pinnata	S
Salvinia cucullata Roxb.	S
Fungi	
Agaricus	
campestris	S
ostreatus	–
igniarius	–
Polyporus	
anthelminticus	–
officinalis	–
Boletus	
crocatus	
Mytitta	
lapidescens	
Auricularia	
sambucina	PS
Lichen	
Parmelia	
kamstchadalis	
perlata	PS
perforata	
Cyperaceae	
Kyllinga	
triceps	S
monocephala	
Fimbrystylis	
junciformis	
Juncellus	
inundatus	
Cyperus	
scariosus	
rotundus	
sculentus	PS
longus	
articulatus	
iria	
Scripus	
grossus	PS
articulatus	
kysoor	
maritimus	

contd.....

Table 4.1 : contd....

Scientific Name	Status	Scientific Name	Status
Poaceae		*Eleusine*	
Oryza		coracena	
sativa		indica	
Hygroryza		**Gymnospermae (Coniferae)**	
aristata		*Cycas*	
Coix		rumphii	D
lachryma-jobi		revoluta	D
Polytoca		*Ephedra*	
barbata		gerardiana	D/C
Zea		*Gnetum*	
mays		scandens	D
Saccharum		*Cupressus*	
officinarum		sempervirens	S
arundinaceum		*Juniperus*	
munja		communis	T
spontaneum		recurva	T
Manisuris		macropoda	T
granularis		*Taxus*	
Vetiveria		baccata	T
zizanioides		*Pinus*	
Amphilophis		longifolia	PS
odorata	PS	gerardiana	–
Cymbopogon		excelsa	PS
jwarancusa		*Cedrus*	
schoenanthus		deodara	PS
nardus		*Abies*	
citratus		webbiana	PS
Heteropogon		**Poaceae**	
contortus		*Phragmites*	
Avena		maxima	PS
fatua		*Dactyloctenium*	
sativa		aegypticum	PS
satvia var. orientalis		*Triticum*	
Desmostachya		aestivum	C
bipinnata	PS	durum	C
Cynodon		spelta	C
dactylon		amyleum	C

contd.....

Table 4.1 : contd....

Scientific Name	Status	Scientific Name	Status
Hordeum		frumentacea	PS
vulgare	C	crus-galli	PS
Paspalum		*Setaria*	
scrobiculatum	PS	italica	C
Pennisetum		plicata	S
spicatum	PS	viridis	S
compressum	PS	*Sorghum*	
Thysanolaena		halepense	C
maxima	PS	vulgare	C
Panicum		*Bambusa*	
miliaceum	PS	arundinacea	C/PS
miliare	PS	*Dendrocalamus*	
antidotale	PS	stricutus	C/PS
Echinochloa			
colona	PS		

LEGEND : C = Cultivated; P.S. = Presently Safe ; S = Sporadic; T = Threatened; D = Not known; S* = Sporadic (gregarious at places)

Chapter Five

Some Potential Drug Plants of India Comments on the list of Dr. Sibakali Bhattacharyya

It is a list that covers the plants of the temperate and tropical regions. The conifers such as Abies, Tsuga, Picea occur in hills at particular heights and gregariously. Most of other species all over are sporadic in occurrence though a few species have gregarious growth over a small area; in legend such species are marked S*.

Most of the species listed by Bhattacharyya are sporadic in occurrence. He has not listed a great number of species, but most of the species listed are common.

On the whole the species listed are facing a crisis. Bhattacharyya's book — "Chiranjib Vanausodhi" has discussed more than 800 diseases; and has listed more than 3000 local names of medicinal plants.

Table 5.1 : Drug Plants — List Prepared by S. Bhattacharyya

Scientific Name	Status	Scientific Name	Status
Abies webbiana	PS	*A. vulgaris*	S
Acacia pennata	PS	*Arundinella nepalensis*	PS
Achyranthes aspera	S	*Aster trinervius*	S
Ageratumconyzoides	S	*Azadirachta indica*	PS/C
Albizzia lebbeck	S	*Bambusa arundinacea*	C/PS
A. procera	PS	*Bauhinia purpurea*	C/PS
Alpinia malaccensis	S	*B. variegata*	C/PS
A. nigra	S	*Bischofia javanica*	PS
Alstonia scholaris	PS	*Bombax ceiba*	PS/C
Aphanamixis polystachya	S	*Caianthe arborea*	PS
Argemone maxicana	S	*Callicarpa macrophylla*	S
Artemisia nilagirica	S	*Candamine hirsuta*	S

contd.....

Table 5.1 contd....

Scientific Name	Status	Scientific Name	Status
Careya arborea	S	*O. nepalensis*	T
Carica papaya	C/PS	*O. stellata*	S
Chukrasia tabularis	PS	*Paederia foetida*	T
Cassia tora	S	*Peperomia pellucida*	S
Cinnamomum tamala	C	*Pinus roxburghii*	PS/C
Cissampelos pareira	S	*Piper betle*	C
Alocasia esculenta	C	*Pongamia pinnata*	PS
Croton caudatus	S	*Populus ciliata*	C
Cynodon dactylon	PS	*Pothos cathcartii*	C
Cyperus rotundus	PS	*P. scandens*	C
Erythrina stricta	C/PS	*Pouzolzia hirta*	S
Eurya japonica		*Rhododendron arboreum*	S
Fagopyrum esculentum	S	*Rungia parviflora*	S
Ficus semialata	S	*Sabia lanceolata*	S
F. fistilosa	S	Saccharum spontaneum	S/PS
F. gibbosa	S	*Salix tetras perma*	S
F. hispida	PS	*Schima wallichii*	PS
F. rumpii	S	*Schleichera oleosa*	C/PS
Garuga pinnata	S	*Spondias pinnata*	PS
Gnaphalium luteo-album	S	*Sterculia villosa*	S
Holmskioldia sanguinea	S	*Stereospermum chelonoides*	S
Homonia riparia	S	*Streblus asper*	S
Imperata cylindrica	PS	*Tamarix dioica*	S
Lannea coromandelica	PS	*Themeda aruninacca*	PS
Leea aequata	S	*Thysanolaena maxima*	PS
L. asiatica	S	*V. negundo*	S
L. inidca	S	*Wendlandia tinctoria*	S
Lygodium flexuosum	S	*Woodfordia fruticosa*	S
Maesa chisia	S	*Wrightia tomentosa*	S
M. indica	S	*Aegle mermelos*	C/PS
M. ramentiaca	S	*Alpinia galanga*	T
Mallotus philippinensis	S	*Alternanthera sessillis*	S
Mangifera indica	C	*Butea parviflora*	PS
Mikania macrantha	S	*Calotropis gigantea*	S
Mimosa pudica	S	*Cassia fistula*	S
Oroxylum indicum	PS/S	*Commelina bengalensis*	S
Osbeckia chinensis	S	*Euphorbia hirta*	S
Osbeckia crinata	S	*Holarrhena pubescens*	PS

contd.....

Table 5.1 contd....

Scientific Name	Status	Scientific Name	Status
Hymenodictyon excelsum	S	*Ardisia crispa*	S
Hyptis suaveolens	S*	*A. Solanacea*	S
Jasminum sambac	C	*Asplenium nidus*	T
Jatropha curcas	S*	*Baliospermum montanum*	T
Justicia adhatosa	S	*Bidens pilosa*	T
Lantana camara	S*	*B. tripatita*	S
Lepidagalhis incurva	S	*Blechnum orientale*	T
Leucus aspera	S	*Boehmeria macrophylla*	S
Ludwigia parviflora	S	*B. nivea*	S
Luvunga scandens	S	*Breynia retusa*	S
Melocanna bambusoides	C/PS	*Buddleja asiafica*	S
Micromelum pubescens	S	*Butea monoperma*	S/PS
Pharagmites karka	S	*Buttneria grandiflora*	S
Phlagocanthus thyrsiflorus	S	*Caesalpinia cucullata*	S
Scoparia dulcis	S	*Casearia vareca*	S
Setaria palmifolia	S	*Canarium bengalense*	T
Dipterocarpus turbinatus	S	*C. strictum*	S
Engelhardtia spicata	S	*Cassia occidentalis*	S
Erythrina variegata	S	*Celastrus paniculata*	S
Ficus cunia	S	*Chasalia euriviflora*	S
F. glomerata	S	*Celtis australia*	S
F. plamata	S	*Chionanthus intermedia*	S
Phoenix sylvestris	S/PS	*Chioranthus elatoir*	S
Quercus lamellosa	S	*Christalla perasitica*	S
Oxalis corniculata	S	*Chroniolaena odorata*	S
Urena lobata	S	*Cinnamomuum bejolghata*	S
S. theaefolia	S	*C. glanduliferum*	S
Tsuga dumosa	S	*Cissus repens*	S
Abutilon indicum	S	*Citrus onlimetta*	C
Acronychia laurifolia	S	*C. medica*	C
A. Pedunculata	S	*Clausena excavata*	S
Aglaea hiernu	S	*Clerodendrurm colebookianum*	S
Ailanthus excelsa	PS	*C. serratum*	S
Altingia excelsa	T	*C. viscosum*	S
Ampelopteris prolifera	S	*Cocculus hirsutus*	T
Anisomeles indica	S	*Combretum acumnatum*	S
Antidesma bunilis	T	*C. pilosum*	S
Aporusa octandra	S	*C. roxburgia*	S

contd.....

Table 5.1 contd....

Scientific Name	Status	Scientific Name	Status
Cordia dictiotoma	S	*Ichnocarpus frutescens*	S
C. joufra	S	*Indigofera tinctoria*	S
C. tiglium	S	*Ipomoea cymosa*	S
Curculigo orchioides	T	*I.purpurea*	S
Dalbergia lanceolaria	S	*Justicia vasiculosa*	S
D. pinnata	S	*Knema linifolia*	S
D. stipalacea	S	*Kaulfussia aesulifolia*	S
Derris indica	S	*Lindenbergia indica*	S
Desmodium gangeticum	S	*Lindera neesiana*	S
D. leterocarpum	S	*L. glutinosa*	S
D. pulchellum	S	*L.Lancifolia*	S
Dichroa febrifuga	S	*Lyonia ovalifolia*	S
Dimocarpus longan		*Mahonia nepaulensis*	T
Dioscorea bulbifera	T	*Melastoma malabathricum*	T
Drymaria cordata	T	*Melodina monogynus*	S
Dryopteris barbegera	T	*Michetiaontana*	
D.odontoloma	S	*Miliusa roxburghiana*	S
Elaeocarpus floribundus	T	*Millettia auriculata*	S
Elsholtzia blande	S	*Phyllanthus*	S
Engelhardtia spicata	S	*Pmaderas patensis*	S
Erioglossum rubiginosum	S	*Picrasma javanica*	T
Evodia fraxinifolia	S	*Pimpinella diversifolia*	S
Flacourtia jangomas	S	*P.sylvatium*	S
Floscopa scandens	S	*P. thomsonii*	S
Forrestia mollissima	S	*Pithecolobium angulatum*	S
Fragaria vesca	S	*P. begeminium*	S
Gardenia campanulata	S	*p. montanum*	S
Geranium nepatense	S	*Pityrogramma calomelonos*	N K
Glochidion lanceolarium	S	*Plantago erosa*	T
Glycosmis arborea	S	*Pogostemon bingalense*	S
Goniothalamus sesquipedalis	S	*P. pubescens*	S
Grewia hirsuta	S	*Polygala urillata*	S
G. laerigata	S	*Polygala longifolia*	S
Haldina cordifolia	S	*Polygonatum mulliflorum*	S
Hedera nepalensis	T	*Polygonum barbatum*	S
Hedyotis seandens	S	*P. chinense*	S
Hovenia dulcis	S	*P.Orientale*	S
Hydnocarpus kurzii	S	*P. rude*	S

contd.....

Table 5.1 contd....

Scientific Name	Status	Scientific Name	Status
P. tomentosum	S	*Phrynium capitatum*	S
Premna latifolia	S	*Phyllanthus reticulatus*	S
Prunus ceraoides	S	*P. urinaria*	S
P. salicina	S	*Rourea minor*	S
P. undulata	S	*Sambucus javanica*	S
Psychotria monticola	S	*Sanicula europea*	S
Pteridium aquilinum	T	*Sapindus mukorossi*	S
Pteris ensiformis	T	*Sapindus varak*	S
Pterospermum acerifolium	S	*Saprosma ternatum*	S
Quercus lamellose	S	*Senecio quinquelobus*	S
Rhamnus nepalensis	S	*S.scandens*	S
Mitragyna rotundifolia	S	*Sida rombifolia*	S
Mucuna imbricata	S	*Smilax glaucophylla*	S
M.pruriens	S	*S. lancaefolia*	S
Mussaenda glabra	S	*S. ocalifolia*	S
Myristica angustifolia	S	*Solanum nigram*	S
Myxopyrum smilacifolium	S	*S.xanthocarpum*	S
Naravelia zeylanica	S	*S. torvum*	S
Nephrolepis cordifolia	S	*Sphenomoris chusane*	S
Ochna squarrosa	S	*Spitanthes paniculata*	S
Olax nana	S	*Stephania japonica*	S
O. scandens	S	*Strobilanthes caleisus*	S
Oldenlandia auriculata	S	*Styrax serrulatum*	S
Olea dioica	S	*Symplocos racemosa*	S
Oleandra wallichii	S	*S. malaccense*	S
Ophiopogon intermedius	S	*Ternsnalia*	S
Ophiorrhiza mungos	S	*T.chabula*	S
Oreocnide integrifolia	S	*T. citrina*	S
Osmanthus fragrans	S	*Tetracer sarmentosa*	T
Olax corniculata	S	*Tetratigma serrulatum*	—
P. scandens	S	*Thaspi avense*	—
Pajanelia longifolia	S	*Tiliacora racemosa*	S
Pandanus furcatus	S	*Toddelia asiatica*	S
Parameria gladulifera	S	*Trachelaspermum lucidum*	S
Pavettia indica	S	*Trema cannabina*	S
Persea bombycina	S	*T. orientalis*	S
Persicaria chinensis	S	*T. nudiflora*	S
Phonebe lanceolata	S	*Triumfetta rhomboidea*	S

contd.....

Table 5.1 contd....

Scientific Name	Status
Tropidia curculioides	S
Uncaria lobata var glauca	S
Valliaris solanacea	S
Ventilago madraspatana	S
Verbascum thapsus	S
Vivurnum colebrookianum	S
V. foetifum	S
Viola biflora	S
Viola patrinii	S
Vitex glabrata	S
Anthoxylum acanthopodium	S
Abroma agusta	S
Acacia dealbata	S
Argyreia nervosa	S
Artemisia indica	S
Bauhinia acuminata	S
B. racemosa	S
Bridelia stipularis	S
Dendrocnide sinuata	S
Derris ferruginea	S
Desmodium elegans	S
Elephantopus scaber	S
Garcinia xanthochymus	S
Garuga gamblei	S
Glochidion lanceolarium	S
Hydrocotyle sibthorpiodes	S
Myrica asculenta	S
Olea diocica	S
Paveta indica	S
Polygala arillata	S
Polygonum perfoliatum	S
Rhus semialata	S
Alpinia nigra	S
Andropogon citratas	C
Anidesma acuminatum	T
Artimisea indica	S
Aesculus assamica	S
Capparis spinosa	S
Hedyotis scandens	S
Lindera candata	S
Diospyros toposia	S
Hydrocotyle nepalensis	S
Michelia doltsopa	S
Mucuna bracteata	S
Sterculia urens	S
Swertia pulchella	S
Vittis peduncularis	S
Zanthoxylum armatum	S
Clerodendrum phlomoides	S
Ligustrum spicatum	S
Dichroa febrifuga	S
Evodia meliaefolia	S
Juniperus pseudosabina	T
Picea smithiane	T
Prunus aeraissoides	T
R. hodgsonii	T
Rhus hooden	S
I	
Acanthopanax trifoliatum	S
Aconitum ferox	T
Aconitum sp.	T
Actcnodapmine angustifolia	S
Adiantum flabellulaum	T
A. pedatum	T
Anisomeles indica	S
Aquilaria malacensis	T
A. platanifolia	
A. lagata	
Caryota urens	T
Cephalotaxus griffthii	T
Chisocheton cumingianus	S
Coptis teeta	T
Cymbidium aloifolium	T
Elaeocarpus	S
Galium aparine	S
Garcinia cowa	S

contd.....

Table 5.1 contd....

Scientific Name	Status	Scientific Name	Status
G. acumilnata	S	*Kydia calicina*	S
G. lancifolia	S	*Lagerstroemia speciosa*	PS
Gymnema acuminatum	T	*Litsaea cubeba*	C
Gynocardia odorata	S	*Melia dubia*	S
Juglans regia	T	*Piper*	C
Panax pseudoginseng	T	*Longum*	C
Podophyllum hexandrum	T	*P. nigram*	C
Polyalthia	T	*P. peepuloides*	C
Simiarum	S	*Plumeria acutifolia*	C
Rhynchostylis retusa	T	*Polyalthia longifolia*	C
Skimmia anquetillia	T	*Potentilla fruticosa*	S
Adiantum caudatum	T	*Piper betle*	C
Scheffera venulosa	S	*Santalum album*	C
Taxus wallichiana	T	*Syzygium cumini*	C
Valeriana wallichi	T	*Tectona grandis*	C
Psidium guava	C	*Terminalia bellerica*	PS
Rhodondron barbatum	T	*T. myriocarpa*	PS
Tamarindus indica	C	*Toona ciliata*	C
Alnus nepaleusis	C	*Zingiber officinale*	C
Cannabis sativa	C	*Embilica officinalis*	C
Anthocephalus chinensis	S	*Momordica charantia*	C
Ocimum sanctum	C	*Nyctanthes arbortistis*	C
Aleurites moluccana	C	*Saraca indica*	C
Allamanda cathartica	C	*Terminalia arjuna*	C
Alocasia fornicata	C	*Agave amernicana*	C
Amomum dealbatum	C	*Camellia sinensis*	C
A. subulatum	C	*Curcuma angustifolia*	C
Carica papaya	C	*Magnolia grandiflora*	C
Colocasia esculenta	C	*Mimosops elengi*	C
D. sissoo	C	*Melia composita*	S
Dillenia indica	S	*Morus laevigata*	C
Gloriosa superba	T	*Mesua ferrea*	C
Gmelina arborea	C	*Michelia baillonii*	—
Hiptage bengalensis	C	*Michelia champaca*	C
Impatiens balasamina	S	*Moringa oleifera*	C
Kleinhovia hospita	S		

Chapter Six

Potential Drug Plants of India Comments on the list of Medicinal Plants Prepared by R.N. Chopra and Indian Council of Medical Research

R.N. Chopra's List

It is not an elaborate one, but a glimpse of the remarks on status of various species may be had. This will show the preponderence of species which are cultivated or threatened of depleted. Only about 23 per cent of the species are presently safe.

This shows the depth of depletion of species.

I.C.M.R's List

The Council described 900 species of plants of which 350 were published in the first volume in 1976 and 550 species were described in 1987.

The Status column shows that cultivated and sporadically occurring speices form a good percentage of species.

Legends has been the same as in other list which are:

S = Sporadic

S* = Sporadic with concentrated regeneration.

C = Cultivated

T = Threateaned

D = Depleted

PS = Presently safe.

Table 6.1 : Chopra and Charka's Work on Potential Drug Plant

Scientific name	Status	Local Name/Parts used
Abroma augusta	S	Abroma bark
Acacia arabica	PS	Indian acacia
Acacia catechu	PS/S	Black catechu
Acalypha indica	S	Indian acalypha
Aconitum chasmanthum		Aconite
Adhatoda vasica	PS	Vasaka
Aegle marmelos	PS	Bael. Bela fructus
Allium sativum	PS	Garlic allium
Alpinia barbadensis	—	Aloes
Alpinia officinarum	S	Alpinia. Galangal.
Alstonia scholaris	S	Alstomia bark.Dita bark
Andrographis paniculata	T	Kalmegh
Anethum graveolens	T	Dill
Anethum sowa	T	Sowa
Arachis hypogea	C	Groundnut
Areca catechu	C	Betel nut
Aristolochia indica	T	Aristolochia
Artemisia marima	S	Santonin. Artemisia santonica
Astragalus strobiliforus	T	Tragacanth
Atropa acuminata	T	Indian belladonna
Atropa belladonna	T	Belladonna
Bacopa monniec (Herpestis monicra)	S	Herpestis santoniea
Berberis aristata	T	Berberis root
Baoerhaavia repens	S	Punarnaba
Brassica integrifolia	C	Sinapis
Brassica juncea	C	Brown mustard
Butea monosperma	PS	Butea seed
Calatropis procera	PS	Calatropis
Camellia sinnsis	PS	Tea plant
Cannabis sativa	PS	Cannabis
Capsicum frutescens	PS	Capsicum
Carica papaya	PS	Papaya
Carum carvi	C	Caraway
Cassia angustifolia	C	Senna
Cassia fistula	PS	Cassia fruit
Centella asiatica	PS	Hydrocotyle
Cephaelise pecacuanha	T	Ipecae

contd.....

Table 6.1 contd....

Scientific name	Status	Local Name/Parts used
Chenopodium album	PS	Wormseed
Chenopodium ambrosioides var. anthelminicum	PS	American worm seed
Chrysanthcemum cinerariafolium	C/D	Pyrethrum
Cinchona ledgeriana C. Succirubra and other species and hybrids	C	Cinchona
Cinnamomum camphora	C	Camphor
Cinnamomum zeylanicum	C	Cinnamon
Cissampelos pareira	D/T	Cissampelos
Citrullus colocynthis	C	Colocynth
Citrus aurantium	C	Bitter-orange ped
Citrus medica var. *limon*	C	Lemon peel
Cocos nucifera.	C	Coconut
Coffea arabica	C	Coffee plant
Colchicum luteum	D/T	Colchicum corm and seed
Coriandrion sativum	C	Coriander
Crocus sativa	C	Saffron
Cuminum cyninum	C	Cumin
Curcuma longa	C	Turmeric
Cymbp pogon flexuosus	PS	Lemon grass
Datura fastuosa	PS	Datura
Datura metal	PS	Datura
Datura stramonium	PS	Datura
Derris ferruginea	PS	Derris. Tuba root
Digitalis lanata	D/T	Digoxin
Digitalis purpurea	D/T	Digitalis leaf
Dryopteris flix mas	D/T	Male fern. Aspidium
Elettaria cardamomum	C	Cardamom fruit
Ephedra gerardiana. E. nebrodensis	D/T	Ephedra
Eucalyptus globulus	PS	Eucalyptus
Eugenia caryophyllus	C	Clove
Eupatoriam ayapana	C	Ayapana
Ferula northex (Ferula factida)	C	Asaioetida
Focniculum vulgare	C	Fennel
Gaultheria fragrantissima	D/T	Wintergreen

contd.....

Table 6.1 contd....

Scientific name	Status	Local Name/Parts used
Glycyrrhiza glabra	D/T	Liquorice
Hemidesmus inducus	D/T	Indian sarsaparilla
Holarrhena antidysenterica	PS	Kurchi bark
Hydnocarpus wightiana	D/T	
Hyoscyamus muticus	D/T	Hyoscyamus
Hyoscyamus niger	D/T	Hyoscyamus leave
Ipomoea hederacca	C	Kaladana
Ipomaa turpethum	C	Turpeth
Juniperus macropoda	D/T	Juniper
Linum usitatissimon	C	Linseed
Lobelia nicotiatafolia	–	Lobelia
Melia azadirachta	PS	Nim
Mentha arvensis	C	Peppermint
Mentha piperita	C	Peppermint
Moringa olcifera	C	Moringa
Myristica fragrans	D/T	Nutmeg
Papaver somniferum	C	Opium
Picrasma quassioides	D/T	Quassia
Picrorhiza Kurrooa	D/T	Picrorhiza
Pimpincll anisum	C	Anise
Pinus excelsa	PS	Colophony
Pinus Khasya	PS	Colophony
Pinus longifolia	PS	Colophony
Piper cubela	C	Betel
Piper ovata	C	Isabgul
Podophyllum hexandrum (Podophyllum emodi)	D/T	Indian podophyllum
Polygala chinensis	S	Indian senega
Prunus amygdaius	S	Almond oil
Psoralea corylifolia	S	Babchi
Ptcrocarpus marsupium	D/T	Kino
Rauwolfia serpentina	D/T	Rauwolfia
Rheum emodi; R. webbianum	D/T	Rhubarb
Ricinus communis	PS	Castor oil
Rosa damascena	CS	Rose
Rosmarimus officinalis	CS	Rosemary
Santichom album	PS	Sandal wood

contd.....

Table 6.1 contd....

Scientific name	Status	Local Name/Parts used
Saraca indica	C	Asoka bark
Saussurea lappa	D/T	Saussurea
Sesamum indicum	C	Sesame oil
Strophantinus Kombi	D/T	Strphanthus
Strycinos nux-domina	D/T	Nux vomica
Swertia chirata	D/T	Chiretta
Terminalia chebula	S	Myrobalam
Thymus vulgaris	C	Thyme
Tinospora cordifolia	D/T	Tinospora
Trachyspermum animi (Carum copticum)	C	Ajowan
Tylophora asthmatica	S	Antamul
Urginea indica	S	Indian squill
Valeriana wallichii and other species	D/T	Valerian
Vernonia anthelmintica	D/T	Vernonia
Vitex peduncularis	D/T	Vitex leaf
Withania somnifera	S/C	Ashwagandha
Zingiber officinale	C	Ginger

Status of Plants (Medicinal) Listed by Indian Council of Medical Research:

The Council described 900 species (350 in their first volume published in 1976 and 550 in their second volume published in 1987)

Table 6.2

Scientific Name	Status	Scientific Name	Status
Abrus precatorius	S	*Carica papaya*	C
Aloe indica	C/T	*Curcuma lonja*	C
Areca catechu	C	*Daucus carota*	C
Azadirachta indica	PS	**Plants with uterine activity**	
Arotabotrys odoratissima	C/S	*Abroma augusta*	S
Butea frondosa	PS	*Anona squamosa*	C
Caesalpinia bondiucella	S	*Aristolochia bracteata*	T
Canscora decussata	S	*Boxbax ceiba*	PS

contd.....

Table 6.2 contd.....

Scientific Name	Status
Cyperus rotundus	PS
Gloriosa superba	T
Gossypium arboreum	S
Acacio farnestnata	PS
Achyranthes aspera	S
Aconitum sp.	T
Acorus calamus	T/C
Aegle marmelos	PS
Artemisia vulgaris	S
Callophyllum inophyllum	S
Calotropis procera	PS
Catharanthus roseus	C
Cestrum diurnum	C
C. nocterrum	C
Cissampelos pareira	S
Cissus quadrangularis	S
Coceulus hirsutus	S
Cocos nucifera	C/PS
Crotalaria juncea	S
Daucus carota	C/PS
Euphorbia reriifolia	S
Evolvulus alsinoides	PS
Plants with Anti-Cancer activity	
Arbus precatorius	S
Albizzia lebbeck	PS
Alstonia scholaris	S/PS
Aphanamixis polystachia	S
Anacardium occidentale	C
Anona squamosa	C
Argemone maxicana	S
Asparagus racemosus	T
Bacopamonnien	S
Calotropis gigantea	S
Cedrus deodara	C
Datura metel	S
Erythrina suberosa	S
Glycosmis arborea	S/PS
Euphorbia hira	S/PS
Gynandropsis pentaphylla	S/PS
Plants with anti-ateros Clerotic activity	
Acorus clamus	C/D
Achyranthes aspera	S/PS
Alpinia galanga	S
Aphanamixis polystachia	S
Anagallis arvensis	S
Argemone maxicana	PS
Aradirachta indica	PS
Capparis decidua	PS
Bryophyllum calicinm	S
Carissa carandus	S
Cassia fistula	P
C. occidentalis	P
Crotalaria juncea	C
Cucurbita maxima	C
C. longa	C
Desmodium gangetium	S
Ervatania coronaria	C/PS
Plants with anti-fungal activity	
Astera cantha longifolia	S
Cassia fistula	PS
Cocos nucifera	PS
Desmodium gangeticum	S
Plants with anti-ateros Clerotic activity	
Aphanamixis polystachia	S
Caesalpinia bonducella	S
C. sepiaria	S
Capparis decidua	S
Cassia auriculata	C
Plants with Anti-bacterial activity	
Acorus calamus	T
Achytranthes aspera	S
Alpinia galanga	S
Eclipta alba	PS

contd.....

Table 6.2 contd.....

Scientific Name	Status
Emblica officinalis	PS
Gmelina arborea	PS
Grewia hirsuta	S
Grewea tilitalia	S
Plants with Anti-protozoal	
Acacia nilotica	PS
Alangium salvifolium	S
Albizzia lebbeck	PS
Asparagus racemosus	T
Atropha belladonna	T
Berberis aristata	S
Toona ciliata	C/PS
Centella asiatia	PS
Cinchona sp.	C/PS
Clerodendrum speciosum	S/PS
Curctuma longifolia	C/PS
Euphorbia hirta	C/PS
Ficus racemosa	S
Celastrus paieulatus	S
Cynodon dactylon	PS
Cyperus nivens	PS
Diospyros preragrina	PS
Dryneria quercifolia	S
Embelia ribes	S
Ficus religiosa	PS
Plants with Insecticidal activity	
Acorus calamus	C/T
Anona squamosa	C
Euphorbia neriifolia	PS
Plants with Diuretic and Anti-urolithiatic activity	
Achyranthes aspera	S/PS
Azadirachta indica	PS
Barleria prionitis	S
Boerhaavia diffusa	PS
Cassia occidentalis	PS
Crataeva nurvala	S
Plants with anti-asthmatic and Antihistamic Activity	
Allium sepa	C/PS
Clerodendrum serratum	S
Curcuma longa	C/PS
Platns with Anti-helmentic activity	
Aegle marmelos	PS
Alangium lamarkii	S
Aphana mixis polystachia	S/T
Butea frondosa	PS
Capparis decidua	S
Datura metel	S
Plants with Respiratory activity	
Alpinia galanga	S
Brassica oleracea	C
Plants with Astringent activity	
Caesalpinia bonducella	S
Capparis decidua	S
Carum coptis	C/PS
Cassia fistula	S/PS

Chapter Seven

Technology on Trade and Commerce, Imports and Exports

This Chapter records various facts on Cultivation, Marketing Economics of cultivation, Trade and Commerce, Conservation, Sustainable harvest of rare medicinal plants, besides presenting lists of major plants required by Indian Pharmaceutical Industries on raw drug materials and a broad outline on export and import of medicinal plants.

A paper cutting (Times of India dt. 12.11.05) on drug market which reveals India's position may be relevant to quote which may be read at the ultimate page of this Chapter.

The author finds it very convenient and relevant to quote the opinion of several technologists on the following issues relating to "Medicinal plants" which are on:

- Cultivation and Marketing.
- Economics of cultivation
- That desert biodiversity on Medicinal plants.
- Trade and Commerce.
- Cultivation and Conservation
- Marketing strategy and Trade.
- Marketing Trade.
- Sustainable harvest.
- Introduction of Nitrages fixing Medicinal plants.
- Rare and less known Medicinal plants.

• N.S. BISHT et al write on " Cultivation and Marketing of Medicinal plants of Pithoragarh". A summary is given below:

The status of collection, cultivation and marketing of medicinal and aromatic plants was studied in Pithoragarh District of Uttaranchal. The primary information was collected as per structural questionnaires from collectors/cultivators belonging to twelve villages spread over two blocks namely Munsiyari and Didihat. The important species being collected were observed to be Jhula, Reetha and Tejpat. The cultivators seem to be growing greater quantities of Atees, Gudhvach, Indrayan, Jambo, Jatamansi, Kalajeera, Kutki, Pashanbhed, Reetha, Sameva and Tejpat. The most favoured market channel was observed to be producer, Middlemen, Trader and Consumer which was being adopted by 50% collectors and 90% cultivators. The producer's share in consumer's rupee in case of collection varied between 45–76. 47% for different species with an average of 56.22%. Similarly the producer's share in consumer's rupee for cultivated species varied between 32.67%– 89% with an average of 60.88%. The paper also discusses the recent changes introduced by Uttaranchal Govt. in marketing of these medicinal and aromatic plant species.

• Mohit Gera et al writes on " Economics of Cultivation" of some medicinal plants a Summary of which is as follows:

Agro technologies for cultivation of a number of medicinal plant species have been developed but large-scale cultivation on farmland is yet to begin. Amongst other causes for this gap, lack of reasonably correct information on econmics of cultivation of these species is one important cause. The economics of cultivation of five medicinal plant species, *viz.*, Kalmegh, Buch, Safed musli, Ashwagandha and Akarkara was studied on farmer's field in Haryana. The net benefits calculated for each species were also subjected to sensitivity analysis in relation to fall in price by 25%, 50%, 75% increase in wage rate by 10%, 20% 30% and increase in rental value of land by 20%, 40%, and 60%. The results showed that maximum net benefits of Rs. 36,140 and Rs. 19,016 per acre could be received by cultivation of safed musli and Kalmegh respectively. The cultivation of Kalmegh and Ashwagandha were observed to be more resilient to the adverse factors of price fall, increase in wage rate and rental value of land, compared to other species. On the basis of initial investment involved and resilience to adverse marked conditions, the cultivation of Kalmegh and Ashwagandha is rcommended for small farmers. The large farmers who can afford greater risk may cultivate Bush and Safed mulsi and Akarkara for higher returns.

• K.K. Chaudhuri et al summarise their work on " Thar Desert Biodiversity for medicinal plant" as follows:

The Thar Desert is the world's seventh largest desert and is most inhospitable eco-region in the Indo-Pacific region. It is spread over in the four states of India, and in Pakistan and cover. an area of about 2,38,700 km^2. There are as many as 157 species of plants with medicinal value. Medicinal value and usefulness of several species is yet to be studied and established. The diversity of the medicinal flora typical of desert ecosystem has immense future prospects. The prominent families to which the majority of the medicinal plants of the arid zone belong are : Fabaceae, Asclepiadaceae, Malvaceae, Acanthaceae, Amaranthaceae, Convolvulaceae, Lamiaceae (Labiatae). The present study is aimed at indexing of all medicinal flora occuring in the Western Rajasthan of Indian desert and their uses for the benefit of the researchers to undertake studies on the prospects and potential of commercial exploitation.

• R.B.S. Rawat et al summarise their work on " Trade and Commerce opportunities of medicinal plant in India".

About 5,000 plant species have been documented for medicinal value and phyto-chemically studied. Of these,1,100 are used in different systems of medicine, 600–700 are used in indigenous industries, but only about 150 have been commercially exploited. Besides domestic use, export potential of these plants is huge and given a quality upgradation of such drugs, competitiveness and globalization is ensured. There is, however need for doing scientific work on their pharmacology, phytochemistry and clinical experiments to develop the export potential fully. Important plant species being utilized at present, their world prices, and other potential species have been listed. Shift towards use of herbal drugs worldwide has been noted. There is good scope in developing this sector. Trade and commerce requirements relating to export and marketing in various foreign marketing e.g., Canada, Hungary, France, UK, USA have been discussed with a view to developing trade in these countries in medicinal products. Various measures, which handicap expansion, have been pointed out. These are : agricultural practices like harvesting and propagation, processing high yield varieties, quality control, marketing, training of personnel, equipment and knowledge about latest advances in technology etc. where efforts need to be focused.

• Dhan Singh et al on " Marketing Strategies and Trade" of medicinal plants write as follows:

Medicinal plants have attracted considerable interest in recent years. Commercial enterprises and local dwellers are regularly exploiting natural heritage of these medicinal plants. There is an urgent need of conservation of these valuable medicinal plants through cultivation. Poor marketing structure in the country is the primary challenge towards its pronotion and cultivation. In Uttaranchal, attention has been given to conservation/cultivation and its open trading system by the government. The present paper highlights the cultivation and open trading aspects of medicinal plants in the State.

• S. Chandola writes on " Some rare and less known medicinal plants of Uttaranchal as follows: (Source : Ind for March 2005)

Valuable species have been removed for so long and so intensively from the wild that they have come to the brink of extinction. The market forces, however have been so strong that substitute have emerged to satisfy the demand, and over time the substitute has assumed the importance of the original drug. The present paper deals with correct identity of Akarkara (Anacyclas pyrethrum), Chirayta (Swertia chirata), Kuth (Saussurea costus), Salam Mishri (Eulophia dabia), Gentiana kurroo has been rediscovered after a lapse of 50 years. It is time now to educate ourselves and to adopt the latest benefits of modern science to retrieve the true herbs species from final annihilation. Serious species Recovery Programmes need to be initiated for the highly threatened plants.

• Manmohan Yadav on " Marketing trade of medicinal plants" writes as follows :

Increasing consumer awarness and preference for herb-based natural products including herbal medicines, has resulted in an unexpected surge in the demand for Medicinal and Aromatic Plants (MAPs) and thus over-exploitation of the medicinal wealth present in the forests. The markets for most Non-Timber Forest Products including NAPs, are highly unorganized and secretive and thus suffer from various market imperfections mainly due to the lack of information about the demand and supply of the products being traded to the disadvantage to the collectors and cultivators and sustainable resource availability. The situation thus calls for the need of information which is reliable and readily avilable to the various stakeholders of medicinal plants trade including the policy makers and implementing agencies. A conceptual framework for such a Market Information System (MIS) with its various stakeholders, type of information, and the output to be generated is discussed in this paper. The suggested MIS is at a conceptual framework and has to be tested in the field at a division level before it is refiend and finalized for its actual implementation. The design and implementation of such MIS on Forest Management Unit (FMU) level can help solve many impediments in the development of the MAPs sector on a sustainable and equitable basis.

• W.D. Thomas et al write on " *Sustainable harvest of medicinal plants*" as follows :

Urbanisation and scientific rediscovery is increasing the demand for herbal products, whose overharvest threatens 30% of the traded species. Sustainable harvest traditions are eroding due to (a) poor prices paid to the raw drug gatherers, (b) competition between, (c) market insecurity, (d) many youth gatherers and (e) vehicular mobility to new collection areas. To revive traditions through equitable benefit sharing, Gram Mooligai (*i.e.*, Village Herbs) Company Limited (GMCL) was established, with gatherer's groups as shareholders in Tamil Nadu state in Southern India. Sustainable Biometric experiments could not continue or help in predicting yield but participant observations and " memory harvesting " revealed that focal species have become rare amongst 25% of their earlier collection areas. Major pressures other that harvest include (a) scarce or untimely rainfall, (b) habitat encroachments and (c) grazing. To earn more better price and sustained business demand, GMCL gathers maximized not quanity but quality through thump rules ("do's and don'ts") of sustainable harvesting practices: (a) appropriate habitat area (b) maturity timing (c) less damaging methods (d) proper post-harvest treatment (e) user (buyer,co-harvester) agreements.

• R.C. Sundriyal weites on "Cultivation and Conservation " of medicinal plants in the Himalayas as follows:

This paper highlights strategy for large scale cultivation and long-term conservation of medicinal plants involving differnt stakeholders in the Himalaya. It is emphasized that the focus of the cultivation could be protection of endangered species and/ or achieving the target of higher income by raising and cultivating market demanding species. The paper discusses selection of potential Medicinal and Aromatic Plants (MAPs) broadly recommended for the region and ways and means of domestication, value addition, product formulation, processing, conservation, infrastructure, and R&D support desired for cultivation of medicinal plants. Designs for marketing and possible funding sources are also given. It is highlighted that if cultivation of MAPs is planned properly, it could emerge as a potential sector to support large number of people with high revenue generation.

• A.K. Parandial et al write on "Introduction of nitrogen fixing medicinal plants" a summary of which follows :

The Garhwal Himalayas are one of the richest floristic zones for the medicinal plants of Indian subcontinent. It provides matchless wealth of more than 300 rare and endangered species of medicinal plants having therapeutic properties. The over-exploitation of these precious material from the Himalayan forest ecosystem over last few decades have not only pushed these towards extinction but also enhanced the problem of soil. erosion, land degradation and loss of biodiversity in the area. Introduction of nitrogen fixing plants may provide an important tool for the ecorestoration attempts in this area. Advocating nitrogen fixing plants having medicinal uses may provide wider acceptability among the local populace from economic as well as soil conservation point of view. The adoptability of indigenous species may be useful for planting and rejuvenating the degraded sites in different altitudinal zones of the Himalayan ecosystem. In the present article an attempt has been made to enumerate the existing nitrogen fixing species of medicinal values at various altitude for the conservation of degraded sited in Garhwal Himalayas.

(Source : Ind. Forester of all opinion : March 2005)

Table : 7.1 : Major Plants Required by Indian Pharmaceutical Industries

Ingredient	Source of supply (August 1999) Cultivation	Wild	Imported	Consumption (Tonnes)
Ajwain (*Carum copticum*)	100%			200
Akkalkadha (*Anacycus phrethrum*)			100%	50
Cardamomum green (*Elettariacardamomum*)	66%		33%	60
Cardamomum big (*Amomum subulatum*)		Assam		NA
Aloes (*Aloe vera*)		Maharashtra Tamil Nadu		200
Amala green (*Emblica officinalis*)	50% South	50%MP/UP Maharashtra		10,000
Anantmool (*Hemidesmus indicus*)		TN,AP		200
Baheda (*Terminalia belerica*)		Maharashtra MP		500
Bhringraj (*Eclipta alba*)		MP, UP, TN, Maharashtra, W. Bengal		500
Brahmi (*Bacopa monnieri*)		Tamil Nadu, West Bengal		700
Kankol (*Piper cubeba*)			150 tonnes	NA
Chitrak (*Plumbago Zeylanica*)		Maharashtra, Tamil Nadu		500
Dalchini (*Cinnamomum Zeylanicum*)			100%	250
Daruhaldi (*Berberis aristata*)		Nainital, Kulu		500
Deodar (*cedrus deodara*)		Nainital, Kulu		200
Gajpippali (*Scindapsus officiale*)			100%	400
Guggul (*Commiphora wightii*)		10% Gujarat, Rajastan	90%	500
Harda (*Terminalia chebula*)		Maharashtra, MP		500
Nutmen/mace (*Myristica fragrans*)	20%	Kerala	80%	500

contd...

Table : 7.1 – *contd...*

Ingredient	Source of supply (August 1999) Cultivation	Wild	Imported	Consumption (Tonnes)
Jambhul beej (*Eugenia jambolana*)		Maharashtra, Gujarat, UP, MP, TN		300
Jatamansi (*Nardostachys grandiflora*)		Nepal, Assam, Kulu		200
Jeshthimadh (*Glycyrrhiza glabra*)			100%	5000
Kadu kutuki (*Picrorrhiza kurroa*)		Kulu (HP), Nepal, Assam		200
Kesar (*Crocus sativa*)		Jammu and Kashmir		5
Clove (*syzygium aromaticum*)	13% Kerala		87%	150
Black pepper (*piper nigrum*)	Kerala			150
Ginger (*Zingiber officinale*)	50% Assam Kerala		50%	500
Ashwagandha (*Withania somnifera*)	50%MP	50% MP		500
Nagkesar (*Mesua ferrea*)				200
Pipramool (*Piper longum*)	AP, Maharashtra			200
Safed musli (*Chlorophytum arundinaceum*)	40% MP MH.	40% MP MH.	20%	25
Shatavri (*Asparagus racemosus*)	50% MP, UP	50% MP, UP		500
Vayvidang (*Embelia ribes*)		Maharashtra, MP		200
Kuchla (*Strychnos Nux vomica*)		Assam, AP, Bihar		1000
Kalmegh (*Andrographis paniculata*)		MP, UP, Bihar		250
Senna	60% TN 20% Gujarat			1000

Source : Mr. Anand Puranik, Chemexcil, Mumbai, Personal communication.

Table 7.2 : Raw Drug Material Imported to India

Botanical Name	Trade Name
Anacycus pyrethrum	Akkalkadha
Asperagus adscendens	Musli white
Atropa belladonna	Belladona
Cinnamomum zeylanica	Dalchini
Commiphora wightii	Guggul
Curcuma zedoaria	Kapurkachri
Elettaria cardamomum	Cardamomum green
Garcinia indica	Kokum
Glycyrrhiza glabra	Jeshthimadhu
Hemidesmus indicus	Sariva
Myristica fragrans	Nutmeg/mace
Piper cubeba	Kankol
Psyllium husk	Isabgol
Rauwolfia serpentina	Sarpgandha
Saussurea lappa	Kusth
Swertia chirata	Chirata
–	Ayurvedic and Unani herbs
–	Saps and extracts of Opium
Azadiracta indica extracts	Extracts of Neem

Source : Planning Commission 2000

Medicinal plant species which are also imported.

1. Glycyrrhiza glabra
2. Pmpinella anisum
3. Thymus vulgaris
4. Opercutina terpertum
5. Cuscuta epithymum
6. Smilax ornata
7. Smilax china
8. Lavandula strechas

(*Source* : Ind. For. March 2005 : R.B.S. Rawat)

Table 7.3 : Top Twenty Medicinal Plants Traded in India in Value Terms

	Trade Name	Botanical Name
1.	Atis	*Aconitum heterophyllum*
2.	Meetha telia/Bachnag	*Aconitum violaceam*
3.	Safed musli*	*Chlorophytum borivillianum*
4.	Guggul*	*Commiphora wighti*
5.	Mamira/Mishmi-bitter	*Coptis teeta*
6.	Salam Panja/Salep	*Dectylorhiza hatagirea*
7.	Vidanga/Baibiranga*	*Embelia ribes*
8.	Nagkesar	*Mesua nagassarium*
9.	Rampatri/Bombay Mace	*Myristica malabarica*
10.	Jatamansi*	*Nardostachys grandiflora*
11.	Gaozaban	*Onosma bracteatum*
12.	Kutki*	*Picrorrhiza kurroa*
13.	Kakra-singi	*Pistacia integrimma*
14.	Sarpa gandha*	*Rauwolfia serpentina Benth*
15.	Manjistha	*Rubia cordifolia*
16.	Chandana/Sandalwood	*Santalum album*
17.	Chobchini gulabi	*Smilax glabra*
18.	Chiraiyta*	*Swertia chirata*
19.	Taggar/Mushkabata	*Valeriana hardwikii*
20.	Banafsha	*Viola pilosa*

*Species selected for detailed market study

(*Source :* Amruth, December 1999)

Table 7.4 : Number of Licensed Pharmacies in Indian System of Medicine

State/Union Territory	Number of licensed pharmacies (April 1999)				Number of licensed pharmacies Holding loan licences*			
	Ayurveda	Unani	Siddha	Total	Ayurveda	Unani	Siddha	Total
Andhra Pradesh	556	222	-	778	4	-	-	4
Assam	39	-	-	39	-	-	-	-
Bihar	228	21	-	249				
Delhi	78	24	-	102	1	-	-	1
Goa	5	-	-	5	1	-	-	1
Gujarat	892	-	-	892	125	-	-	125
Haryana	210	3	-	213	-	-	-	-
Himachal Pradesh	54	-	-	54	-	-	-	-
Jammu & Kashmir	10	-	-	10	-	-	-	-
Karnataka	241	-	-	241	20	-	-	20
Kerala	962	-	-	962	9	-	-	9
Madhya Pradesh	225	12	-	237	11	-	-	11
Maharashtra	757	-	-	757	243	-	-	243
Orissa	160	-	-	160	-	-	-	-
Punjab	149	-	-	149	2	-	-	2
Rajasthan	388	4	-	392	-	-	-	-
Tamilnadu	218	8	323	549	17	3	6	26
Tripura	1	-	-	1	-	-	-	-
Uttar Pradesh	2,575	217	-	2,792	2	-	-	2
West Bengal	620	22	-	642	21	-	-	21
Chandigarh	2	-	-	2	-	-	-	-
Dadra and Nagar Haveli	10	-	-	10	-	-	-	-
Deman and Due	1	-	-	1	1	-	-	1
Pondicherry	24	-	94	118	1	-	-	1
Total	**8,405**	**533**	**417**	**9,355**	**458**	**3**	**6**	**467**

Source : Website of Department of Indian Systems of Medicine and Homeopathy.

Table 7.5 : Plants Considered in Market Study by TRAFFIC (India)

	Trade Name	Botanical Name
1	Atis	Aconitum heterophyllum
2	Agar	Aconitum melasensis
3	Daruhaldi	Berberis spp.
4	Safed musli*	Chlorophytum borivillianum
5	Guggul*	Commiphora wightii
6.	Mameera	Coptis teeta
7	Salabmisri	Orchis latifolia
8	Kutki*	Gentiana kurroa
9	Kalihari	Gloriosa superba
10	Kapoorkachri	H. spicatum
11	Pushkarmool	Inula racemosa
12	Jatamansi*	Nardostachys grandiflora
13	Kutki*	Picrorrhiza kurroa
14	Vankakadi	Picrorhiza hexandrum
15	Lal chandan	Pterocarpus santalinus
16	Sarpgandha	Rauwolfia serpentina
17	Kuth	Sassurea lappa
18	Chirayata	Swertia chirata
19	Talispatra	Taxus wallichiana

(*Source* : Mr. Manoj Mishra, TRAFFIC-India: personal communication)

Notification prohibiting the export of medicinal plants

Ministry of Commerce, Government of India

Notification No. 24(RE-98)/1997–2002

New Delhi, Dated 14 October 1998

S.O.(E) Attention is invited to para 4 of Notification No. 2(RE098)/1997-2002 dated the 13th April, 1998 relating to export of plants, plant portion and their derivatives and extracts obtained from the wild.

The export of under-mentioned 29 plants, plant portions and their derivatives and extracts as such obtained from the wild except the formulations* made therefrom, is prohibited.

Cycas beddomei (Beddomes cycad)
Vanda coerulea (Blue vanda)
Saussurea sp.
Paphiopedilium species (Ladies slipper orchids)
Nepenthes khasiana (Pitcher plant)
Renathera imschootiana (Red Vanda)
Rauwolfia serpentina (sarpagandha)
Ceropegia species
Frerea indica (Shindal Mankundi)
Podophyllum hexandrum (emodi) (Indian Podophyllum)
Cyathea species (Tree Ferns)
Cycadaceae species
Dioscorea deltoidea (Elephant's foot)
Euphorbia species (Euphoribias)
Orchidaceae species (Orchids)
Pterocarpus santalinus (Red Sanders)
Taxus wallichiana (Common Yew or Birmi leaves)
Aquilaria malaccensis (Agarwood)
Aconitum species
Coptis teeta
Coscinium Fenestratum (Calumba wood)
Dactylorhiza hatagirea
Gentiana kurroo (Kuru, kutki)
Gentum species
Kamphergia galenga
Nardostachys grandiflora
Panax pseudoginseng
Picrorrhiza kurroa
Swertia chirata (Charayatah)

Major medicinal plant species exported from India

Plant name	Plant part exported
Plantago ovata	Seed and husk
Cassia angustifolia	Leaf and pod
Rheum australe	Rhizome
Inula racemosa	Rhizome
Rauwolfia serpentina	Roots
Hedychium spictatum	Rhizome
Zingiber officinale	Rhizome
Colchium luteum	Rhizome and seed
Acorus calamus	Rhizome
Adhatoda vasica	Whole plant
Juglans regia	Bark
Punica granatum	Flower, root, bark
Barbris aristata	Root
Juniperus communis	Fruit
J. macropoda	Fruit
Heracleum candicans	Rhizome
Picrorrhiza kurroa	Root
Aconitum species	Root
Saussurea lappa	Rhizome
Swertia chirata	Whole plant
Podophyllum emodi	Rhizome
Valeriana wallichi	Rhizome

(*Source* : Handa 1992.)

- Medicinal plants having export potential

  Palntago ovata (Isabgul), *Catharenthus rosea* (SadaBahar), *Aloe vera, A. barbadensis* (Kumari), *Garcinia cambogia* (Kokum), *Gymnema sylvestre* (Madhunasini), *Ocimum sanctum* (Tulsi), *Picrorhiza kurroa* (Kutki), *Phyllanthus nirula; Pamarus* (Bhumianalaki), *Holarrhena pubescens* (Kutaja).

- Species banned for export.

  The Director Genral, Foreign Trade (DGFT) under the foregn Trade Development Act–1992 prohibits the export of the following :

Aconitum sp., Aquilaria malaccensis, Ceropegia sp., Coptis teeta, Consciniuim fenestratum, Cyathea sp., Cycas beddomei, Dactylorhiza hatagirea, Dioscorea deltaidea, Frerea indica, Gentiana kurroa, Gnetum sp., Kaempferia galanga, Nardostachys jatamansi, Nepenthes khasiana, Orchidaceae sp., Panax pseudoginseng, Paphiopedilium sp., Picrorhiza kurroa, Podophyllum hexadrum, Pterocarpus santalinus, Rauwolfia serpentina, Renauthera imschootiara, Saussurea costus, Swertia chirata, Taxus wallichiana, Vanda coerulea.

- Collection strictly banned in Uttaranchal

  Paris polyphylla, Nardostachys jatamansi, Microstylis muscifera, Lilium polyphyllum, Habenaria intermedia, H. edgeworthii, Gentiana kurroa, Frillilaria royleu, Eulophia dabia, Dactylorhiza hatagirea, Arenebia benthamii, Juglans regia (Akhrot), Aconitum heterophyllum (Alees), *Acorus calamus* (Bach), *Podophyllum hexandrum* (Bankakri), *Viola serpens* (Banafsha), *Swertia chirata* (Chirayata), *Jurinea sp.* (Dhop jad), *Dioscorea sp., Rheum emodi* (Dolis), *Angelica glauca* (Gandrayan), *Stephania glabra* (Ginjari), *Morechella esculenta, Myrica esculenta* (Kaiphal), *Lilium heterophylum* (Kakolisir), *Berberis aristata* (Kilmora), *Picrorhiza kurroa* (Kutki), *Malaxis muscifera* (Lahusunia), *Polygonatum cirrhifolium* (Mahamaida), *Aconitum ferox* (Meetha), *Skimmia laurata* (Nairpati), *Didymopearpus pedicellate* (Patharlong), *Onosma sp.* (Ratanyot), *Hebenaria intermeda* (Ridhi-vridhi), *Eulophia campestris* (Salam mishri), *Velleriana wallichi* (Sameva), *Ephedra gerardiana* (Somlata), *Taxus baccata* (Thuer).

Table 7.6 : Medicinal Plants in Short Supply

Botanical Name	Common Name	Quantity required (Tonnes/ annum)	Period of shortage of supply (Years)
	Ashtvarga	0.095	23
Acacia catechu	Khair chhal	2.40	5
Aconitum heterophyllum	ativisha	0.55	15
Alpinia galanga	Kosthakulinjan	0.22	4
Aquilaria agallocha	Krishnageru	0.17	12
Artemisia maritima	Kirmani ova	0.33	5
Artocarpus heterophyllus	Phanas ambe	0.055	5
Baliospermum montanum	Dantimool	0.32	3
Berberis aristata	Daruhaldi	2.70	6
Cedrus deodara	Devdar	2.20	10
Commiphora wightii	Guggul	2.30	5
Convolvulus arvensis	Harenvel	0.156	6
Curculigo orchioides	Kalimusli	2.25	4
Curcuma zedoaria	Kapurkachri	0.225	5
Dioscorea bulbifera	Dukkarkand	0.175	7
Embelia ribes	Vaividang	3.40	3
Gentiana kurroo	Triman	0.22	4
Hemidesmus indicus	Chavak	1.20	3
Indula racemosa	Pokharmool	0.65	6
Mallotus phillippinisis	Kapila	0.155	12
Mesua ferrea	Nagkesar kala	0.65	6
Myrica esculenta	Kaiphal	0.225	5
Myristica fragrans	Jaiphal	0.33	3
Nardostachys grandiflora	Jatamansi	0.66	5
Nelumbo nucifera	Kamalphool	0.31	8
Picrorrhiza kurroa	Kutaki	1.55	5
Piper cubeba	Kankol	0.335	5
Piper longum	Pippali	1.25	5
Piper longum	Pippalmool	0.85	5
Piper nigrum	Shvet miri	0.09	13
Pistacia chinesis	Kakdashingi	0.45	10
Plumbago zeylanica	Chitrak lal	3.50	5
Pterocarpus santalinum	Raktchandan	1.025	18
Rubia cordifolia	Manjishtha	1.15	4
Saraca indica	Ashok chhal	6.80	4
Saussurea lappa	Koshtha	0.43	5
Smilax china	Chopchini	0.55	5

contd...

Table 7.6 – *contd...*

Botanical Name	Common Name	Quantity required (Tonnes/ annum)	Period of shortage of supply (Years)
Solanum indicum	Motiringani	1.15	5
Swertia chirata	Kirata	2.50	7
Tecoma undulata	Raktrida	0.30	6
Valeriana wallichii	Tagar	0.275	5
Vetiveria zinzanioides	Vala	1.15	4
Wagatia spicata	Vakeri bhate	0.12	4
Wrightia tinctoria	Andrajava	0.418	5

The data refers to Sandu Brothers, Mumbai, Who are manufacturers

Source: Planning Commission 2000

Table 7.7 : Estimation of Raw Material Demand of Threatened Species

Notified species	Annual consumption (kg.)
Aconitum species	11,671
Acorus species	109,760
Aquilaria malaccensis	48,599
Artenesia species	795
Atropa species	1,629
Aristolochia species	6,459
Colchicum luteum	1,.637
Coscinum fenestratum	3,300
Costus speciosus	2,186
Commiphora wightii	68,383
Didymocarpus pedicellata	1,527
Ephedra species	84
Gloriosa superba	1,414
Hyoscyamus niger	1,055
Hydnocarpus species	72,645
Orchidaceae species	1,438
Pterocarpus santalinus	15,873
Nardostachys grandiflora	14,228
Rheum emodi	235
Rauwolfia serpentina	11,083
Strychnos potatorum	23,425
Swertia chirata	23,185
Taxus baccata	23,636

China may Pip India in Race for US Drug Market

China is fast catching up and may soon overtake India in the number of DMFs (drug master fillings) to the US Food and Drug Administration. The number of DMF fillings being made by Chinese players has started picking up, and is growing at a fast pace.

DMFs refer to the submission of data that includes technical, clinical and safety information about an API (Active Pharmaceutical Ingredient) to US FDA, and are followed by the abbreviated new drug application (ANDA) filling.

The data submitted by the companies is reviewed by US FDA and only after the companies get an approval from the authority can their active pharmaceutical ingradients be marketed in US. In the first half of 2005, China's share in the total new filling to US FDA was nearly 9% with 37 new filling druging the period, up from a share of 8.1% for last year (49 filings).

The pace at which the filling have been growing over the past two years is something to look out for, industry analysts stated. It will soon be filing for ANDA or abbreviated new drug application in US FDA, which is the next step after DMFs, they added.

Commenting on the pace of DMFs by China, Dr. Swati Piramal, director (strategic alliances & communications), Nicholas piramal India said:"Yes it is fast and something to watch out for ".

There has also been a surge in the number of fillings by Indian Companies, with India's share of new filling at around 33% in the first half of this year. Last year, the share was 32.8% with 198 new fillings. Over the last couple of years, serveral second/third tier companies have aggressively scaled up their ANDA/DMF fillings the US market.

While China may yet trail behind India on issues of quality, chemistry skills and regulatory issues, analysts believe that the gap is likely to narrow going forward, especially in basic products that do not involve advanced chemistry knowledge, experts pointed out.

Meanwhile, there is a section within the industry which felt otherwise. Ramesh Adige, executive director of Ranbaxy Laboratories, said:"The Indian pharma industry is miles ahead in its new fillings and China is just about getting there. I believe this gap will sustain for another five years."

Nimish Mehta, assistant VP (research) Edelweiss Capital, felt that though China has shown an increase in DMF fillings, it is not very major. "This is because Indian companies are targeting companies whose patents are expiring so these are relevant, which may not be the case in the former country," he said.

Major Indian Plant entering Global Commerce

Scientific Name	Scientific Name
Acorus calamus	*Panax* sp.
Agave sisala'na	*Rheum* sp.
Aloe vera	*Saussaria* sp.
Aloe sp.	*Dicorhiza* sp.
Ammi majas	
A. visnaga	**The following have pretty high demand**
Atropa acuminata	
Atropa belladonna	*Tinospora cordifolia*
Berberis vulgaris	*Pulchea macniosa*
Catharanthus roseus	*Asparagus racemosus*
Datura metel	*Centella asiatica*
D. stramonium	*Cassia angustifolia*
Digitalis sp.	*Terminalia chebula*
Ephedra sp.	*Aloe vera*
Derboisia sp.	*Withania sommifera*
Glycyrriliza glabra	*Acontium balfourin*
Hyoscyanus niger	*A.dienorrhizum*
Papavar sommiferum	*A. falconeri*
Rawalfia serpentina	*A. ferox*
R. tetraphylla	*A. heterophyllum*
Rheum emodi	*A. violaceum*
R. officinalis	*Acorus calamus*
Swertia chirata	*Aguilaria malacensis*
Taxus baceata	*Angelica glauca*
Wrginea maritima	*Atropa acuminata*
Valeriana officinalis	*Berberis khahmiriana*
V. wallichii	*Swertia chirata*
Curcuma sp.	*Saussaria* sp.
Gentiana sp.	*Polygonatum* sp.
Inula sp.	*Rauwalfia* sp.
Ilex khasianum	*Dioscorea* sp.
Podophyllum sp.	
Coptis teeta	**Vulnerable species**
Angelica glance	*Berberis* sp.
A. benthami	*Curculigo* sp.
	Hedycium sp.
Endangered species	*Paeoni* sp.
Beberis sp.	*Rheum* sp.
Herecleum sp.	*Clerodendrum* sp.

Chapter Eight
Status Assessment of Tree Flora

There is definitely some discrepancy about the number. But herbs definitely surpass the number of trees and also in their quality and efficacy; it has been mentioned that herbs have at least four times the number of tree species in their occurrence. Similar study on herb and shrub flora should be the first step to assess their status.

In this chapter the author has picked up results of inventory of trees of some states which have commercial values mostly (not specifically having medicinal properties). The purpose is to record the depth of work of the foresters on the status of tree species. In this report the occurrence of trees per hectare only has also been mentioned. One can assess the exact status of each species having known the area of occurrence, type of forests and the area coverage.

For example the occurrence and status of some trees having medicinal values can be assessed from the data presented.

Quantitative Status (density per hectare) Assessment of Tree Flora made by Forest Survey of India, (FSI):

While no statistical-base quantitative assessment of herb/shrub/climber flora which form 90 per cent of medicinal of India, has been done on a wide and acceptable scale, the FSI, has already done extensive and intensive work on such assessment of tree flora all over the country for the last 35 years.

On the basis of this survey it is possible to evaluate the status of a species in a particular area. Such figures also show the valverability/rarely/or the exact status of a species.

Only a few examples are presented as floras (figures relates to 1970s and 1980s).

West Bengal:

a) Darjeeling Hills of West Bengal:

A token survey was carried out in Singalila area and following species were listed:

Rhododendron sp.	67.40 stems per ha.
Lithocarpus pachyphylla	24.00 stems per ha.
Tsuga dumosa	20.50 stems per ha.
Castanopsis sp.	17.70 stems per ha.
Abies densa	14.50 stems per ha.
Lithocarpus lineata	12.50 stems per ha.
Acer campbellii	5.00 stems per ha.

Other species are – *Alnus nepalensis, Simingtonia populnea, Cinnamomum* sp. etc.

b) Kalimpong Hills of West Bengal:

According to a token survey carried out over small area the following species were listed.

Machilus gammieana	22stems per ha.
Lithocarpus lineata	15stems per ha.
Alchimondra cathcartii	15 stems per ha.
Lithocarpus pachyphylla	10 stems per ha.
Nyssa javonica	9 stems per ha.
Castanopsis sp.	8 stems per ha.
Acer campbellii	8 stems per ha.
Cinnamomum sp.	8 stems per ha.
Lithocarpus lamellosa	7 stems per ha.

Overall density per hectare is 275 trees.

c) Purulia District of West Bengal:

The forests of purulia district grow over barren red lateritic soil, 80% of the crop are of 5 cm. to 9 cm. diameter

Shorea robusta	232 stems per ha.
Terminalia tomentosa	23.750 stems per ha.

Semecarpus anacardium	17.000 stems per ha.
Mallotus philippinensis	15.750 stems per ha.
Diospyros sp.	36.5867.40 stems per ha.
Holarrhena sp.	
Pterocarpus marsupium	

Miscellaneous Stratum:

Holarrhena antidysentrica	41.059
Anogeissus latifolia	13.16
Bauhinia purpurea	9.73
Cochlospermum gossypium	10.00
Lannea coromendelica	17.89
Mallotus philippinensis	36.58
Terminalia tomentosa	51.06
Shorea robusta	43.32
Emblica officinalis	13.16
Pterocarpus marsupium	10.26
Gmelina arborea	8.94
Saccopetalum tomentosa	7.30
Croton oblongifolius	6.84
Buchanania lanzan	5.79

Sikkim State

Figures of south Western Sikkim are only available:

Prime species are: Symplocos theifolia, (Frequency occurrence is 90, and stems per ha. is 69); has the maximum density, this species is of immense medicinal value of importance (Frequency given in the brackets.).

Species are *Castanopsis* sp. (Fr. Oct.90, and 13 Stems/ha.) *Machilus odorattissima* (F-66 and 8/ha). *Eurya Japonica* (F-80 8/ha.) *Macaranga* sp. (F-28,7/ha.), *Lithocarpus lamellosa* (Quercus). *Acer campbellii* (F-50, 8/ha.), *Litsaea* sp. (F-90,3/ha.), *Machilus gammeana* (F-70).

Bihar State (Sal bearing tract):

Survey area 53,000 km. In Chhotanagpur hills.

Forest type4 : 5 BC, (C)-Northern Tropical Dry peninsular Sal forests. Stems per hectare:

Sal (poor quality)	450 Stems/ha. (In Sal Forests)
Boswellia serrata	22/ha. (In Sal Forests).
	111/ha. (In Boswellia forests).
Acacia catechu	132/ha. (In Acacia forests).
Miscellaneous species	63/ha.

Species are Terminalia tomentosa, Anogeissus latifolia, Lannea coromendelica, Buchanania lanzan, Syzygium cumini, ficus species, Diospyros melanoxylon, Ougenia oojeinensis, Boxbax ceiba, Madhuca latifolia, lagerstroemia parviflora, Adina cordifolia, Butea monosperma, Schleighera oleosa and pterocarpus marsupium.

Melghat Forests of Maharashtra State:

These forests are close to the forests of Akola, Buldhana of Nimar and Belul districts of Madhya Pradesh.

The forests fall under southern tropical Dry Deciduous Teak type (5A/C1b) and southern tropical Dry Mixed Deciduous forest type (5A/C_3) as per Champion and Seth's forest Types of India. Teak forests occur extensively with Dendrocalamus as understorey, where Dendrocalamus is absent Lantana and Cymbopogon martini and other tall grasses from the ground vegetation.

Melghat Tiger Reserve:

Quantitative survey carried out by Forest Survey of India and Tiger Project authorities, shows the occurrence of 277 trees per. ha. dominant sp. as follows:

High Frequency (%) species are:

Tectona	10
Anogeissus	70
Lagerstoemia	70
Wrightia	6
Emblica	60

Dalbergia	10
Flacourtia	60
Boswellia	60
Acacia catechu	70
Bamboo	50

Southern Tropical Dry Deciduous forests: (Average ht.20.27 m.). Some species are:

Tectona grandis	Lannea coromendelica
Terminalia tomentosa	Qugenia oojeinensis
Wrightia tinctoria	Cassia fistula
Bauhinia racemosa	Terminalia belerica
Mitragyna parviflora	Schleichera oleosa
Casaeria elliptica	Zizyphus xylopyra
Garuga pinnata	Boxbax ceiba
Flacourtia indica	Lagerstroemia parviflora
Emblica officinalis	Anogeissus latifolia
Boswellia serrata	Acacia sundra
Miliusa tomentosa	Terminalia Arjuna
Madhuca latifolia	Adina cordifolia
Albizzia procera	Chloroxylon swietenia
Grewia lilaefolia	Diospyros melanoxylon
Bridelia retusa	Sherbera swietenoides

Some Species of Woody climbers are:

Butea superba	Millettia auriculata
Celastrus paniculatus	Zizphus rugosa
Capparis zeylanica	Cayratia auriculata
C. trifolia	Dregea vollubilis
Hemidesmus indicus	Mucuna prurita
Bauhinia vahlii	Pueraria tuberosa
Combretum ovalifolium	Acacia pinnata
Cissampelos pariera	Ampelocissus latifolia
Cissia repanda	Canavalia gladiata
Crytolepts buchanani	Argyreia sericea

Some herbs are (of Melghat):

Cleome viscosa	Triumfetta pentrandra
T. rhombcidea	Bioshytum sensetivum
Cardiopermum helicacabum	Polygala elongata
Hybanthus aspaspermus	Bidens biternata
Spoubia delphinifolia	Hemigraphis latebrosa
Rungia elegans	Lavendula burmanii
Galinsoga parviflora	Cryptolepis buchanani
Ipomoea quomoclii	Argyreia sericea
I. pestigris	J. obscura
Crinum sp.	Euphorbia prunifolia
Urena lobata	Chlorophyum tuberosum
I. laxiflorum	Desmodium gangeticum
Pogostemon bengalensis	Orthosiphon rubicundus
S. alba	Sida acuta
Leea edgworthii	S. orientalis
P. monoica	Pinpinela lateriflora
S. stricta	Spermacoca hispida
Lepidagathis cristatia	Dipteracanthus patulus
Conyza stricta	L. trinervis
Hemidesmus indicus	Tridex procumbens
Achyranthes aspera	Trichodesma indicum
Phyllanthus debilis	Celocia argentea
Habenaria platyphylla	P. virgatus
D. hispida	Dioscorea belophylla
D. pentaphylla	D. bulbifera
Plectranthus rugosus	Uraria refescens
Anisomeles indica.	

Some herbs growing in Agricultural lands (Melghat):

Argemone mexicana	Sida acuta
Cassia fumila	Spermacoce stricta

Ageratum conyzoides
Vernonia sp.
Solanum migrum
Leucas cephalotes
Phyllanthus sp.
Hibiscus lobatus
Aminania baccifera
Caesulia axillaries
Mollugo pentaphylla
Striga angustifolia
Achyranthes sp.
Lannea procumbens
Echinostemna hyssopifolium
Justicia simplex
Celocia sp.
Merdania nudiflora
Psoralia corylifolia
Oldenlandia corymbosa
Tridax sp.
Merremia gangetica
lepidagathis cristata
Euphorbia sp.

Some Shrubs (Melghat) are :

Lantana camara, sida-5 sp., *Grewia-5* sp., *Alysicarpus-8* sp., *Bhima-6* sp. *Justicia-5* sp. *Leucas-6* sp., *Commelina-5* sp. *Commelina-5* sp., *Flimbristylis-6* sp., *eragrostis-9* sp. *Desmodium-5* sp., *Indigofera-8* sp., *Bauhinia-5* sp. *Cassia-7* sp., *Ipomoea-10* sp., *Euphorbia-6* sp., *Ficus-7* sp., *Cyperus-7* sp. *Brachiaria-5* sp.

Echinochloa frumentacea
P.scrobiculatum
Coix aquatica
Eragrostis ciliaris
Setaria intermedia
Apluda mutica
Chloris dolichostachys
Eleusine coracana
Chloris mirgata
Echinochloa colona
Sporobolus diander
Schmia nervosum
Paspalidium flavidum
Brachiaria enciformis
Eleusine aegyptica
E. tenalla
S. pallidaefusca
Themeda quadrivalvis
Panicum miliare
Setaria italica
Cynodon dactylon
Iseilema laxum
Heteropogon contortus
Capillipidium assimele

Some Trees of Mewasi, Maharastra:

The forests of Mewasi fall under southern Dry Deciduous forests, according to Champion and Seth's Classification of forest types; Where Tectona grandis predominates. But there are wide areas, where miscellaneous crop predominates with a small proportion of Teak.

Stems per ha. Of different species, in the 3(three) categories of forests are latulated below:

Table 8.1

Species	No. of stems per hectare		
	Teak & Misc Forests	Misc. Forests	Teak Forests
Tectona grandis	24.6	5.1	36.8
Terminalia crenulata	9.0	10.5	6.3
Pterocarpus marsupium	2.3	0.5	-
Dalbergia latifolia	4.3	1.4	3.6
Qugenia oodienensis	2.6	3.1	2.13
Diospyros melanoxylon	3.3	0.5	3
Bridelia refusa	5.3	2.8	1.0
Anogeissus latifolia	11.6	3.1	8.5
Lagerstroemia parviflora	5.9	2.5	3.6
Boswellia serrata	5.3	3.7	4.2
Lannea coromendelica	12.6	4.2	6.3
Garuga pinnata	5.0	3.7	0.8
Schleichera oleosa	4.3	0.5	0.4
Terminalia belerica	1.9	3.1	0.4
Adina cordifolia	2.6	0.5	1.0
Mitragyna parmiflora	1.3	2.5	0.4
Albizzia procera	0.6	0.8	-
Butea monosperma	15.0	16.5	6.3

Other Inventory Results in Arunachal Pradesh

The forests have been thoroughly surveyed in the eastern and western districts. The survey shows a rich assemblage of species of modferate density. The area, however, has been suffering depletion due to shifting cultivation and large scale Army settlements.

(a) East Kameng District:

Broad-leaved stratum: 168 econimically important species have been indentified (229 trees per hectare). Occurrence of species per hectre are furnished below:

Nepal (Western)

Table 8.2 : Stems/ha. of Important Species in different Forest types

Species	Forest types					
	Chir	KIL Deodar	Firspruce Tsuga	Conifers mixed with H.W.	High level broad leaved	Low level broad leaved
Chir (*Pinus oxburghii*)	109.136	-	-	-	-	-
Kail (*Pinus allichiana*)	0.021	214.228	15.015	31.672	1.975	-
Deodar (*Cedrus deodara*)	-	2.532	-	-	-	-
Fir (*Abies pindrow*)	-	4.630	132.862	33.183	3.094	-
Spruce (*picea smithyana*)	-	6.622	18.223	13.246	0.399	-
Tsuga (*Tusga dumosa*)	-	4.630	6.506	8.450	0.211	-
Thuner (*Taxus baccata*)	-	-	19.165	5.325	1.824	-
Cyprus (*surai*)(*Cupressus toralosa*)	-	5.6958	34.001	-	-	-
Allo oaks (*Quercus sp.*)	6.040	35.254	24.807	43.335	125.997	0.693
Bhojpatar (*Betula sp.*)	-	9.461	30.381	22.808	8.970	-
Kanjal (*Bischofia javanica*)	-	-	3.638	56.584	3.232	-
Sai (*shorea robusta*)	5.298	-	-	23.644	0.080	178.580
Sain (*Terminalia sp.*)	-	-	-	2.060	-	21.631
Dhawra	-	-	-	0.501	-	25.638
Up land hardwoods	17.882	16.992	56.248	45.651	237.572	10.718
Low land hardwoods	11.623	-	-	4.786	3.130	166.784
Total conifers	109.157	238.300	225.772	114.429	8.988	8.862
Total broad leaved	**40.843**	**61.707**	**115.074**	**149.369**	**379.001**	**404.044**
Grand total	**150.000**	**300.007**	**340.846**	**263.798**	**387.989**	**412.906**

Castanopsis indica\ Castanopsis sp.-14.108, Altingia excelsa-7.233; Ficus dp.-6.566; Quercus sp.-6.464; Sygygium sp.-6.361; Alnus nepalensis-4.668; Canarium sp. 4.66; Macaranga indica-4.412; Dysoxylum binecteriferium 4.104; Amoora sp.-2.719; Canarium resiniferum-2.616; Magnolia sp. 2,462; Trema orientalis-2.360; Engelhardtia spicata-2.155; Michelia, sp. -4.925.

Bamboo startum: 9 species are prominent *e.g.,*- Castanopsis, Altigia, Bischofia, Macarang, Sterculia, Quercus Aostonia, Michelia, Sehia (Per hectare occurrence of each species is between 2 to 3).

Altogether 168 species of trees have been recorded alongwith thier occurrence per hectare. Of these species most widely occurring trees have been depicted. The occurrence of ficus, Gmelina, Glochidion, Artocarpus, Antidesma, Cordia, Dillenia, Dysoxylum, Erythrina, Eugenia, Terminalia species and species of several other trees indicate a bio-ecological bonanza for bird and animal fauna to survive.

(b) West Kemeng District:

The surveyed area represents many tree species. About 39.438 stems per hectre have been left unidentified. The area commands a beautiful landscape having about 35.756 stems of Rhododendron per hectare. Besides, the occurrence of Terminalias (3.779), Eurya (5.394), Acer, Betula, Amoora, Canarium, Aquillaria, Castanopsis, Ehretia, Carya, Eriobotrya, Eugenia, Feronia, Ficus Gmelina, Juglans, Machilus(5.943), Magnolia, Michelia (3.325), Morus, Myristica, Spondius etc. species make the area biologically very interesting.

Boadcleaved stratum: (Number of stems per hectare is 223. 678).

Phoebe sp.	3.359	Quercus griffithii	13.631
Acer sp.	2.972	Quercus lineata	4.490
Schima wallichii	2.972	Quercus sp.	16.892
Sygygium cumini	2.907	Castanopsis indica	7.623
Dysoxylum binecteriferum	2.132	Abies pindrow	6.718
		Alnus nepalensis	8.491
Eugenia sp.	2.100	Eurya japonica	5.394
Terminalia sp.	3.773	Machilus odoratissima	3.585

Tripura State

Most important families are:

Fabaceae	44 genera, 94 species
Poaiceae	49 genera, 91 species
Rubiaceae	38 genera, 78 species
Euphorbiaceae	35 genera, 69 species

Principal tree species (according to percentage) are:

Albizzia procera	7.3
Vitex peduncularis	5.2
Bombax ceiba	1.5
Schima wallichii	4.4
Dillenia pentagyna	0.4
Terninalia belerica	1.4
Gmelina arborea	2.9
Lannea coromendelica	4.4
Shorea robusta	4.5
Artocarpus chaplasa	1.5
Toona ciliata	6.0

Assam State

a) Evergreen forests:

Ammora wallichii	2.40
Dipterocarpus macrocarpus	29.50
Shorea assamica	10.32
Mesua ferrea	13.24
Mangolio pterocarpa	7.00
Cinnamomum glanduliferum	2.17
Vatica lancaefolia	4.70
Terminalia myriocapra	1.60
T. chebula	4.60
t. belerica	1.08

b) Evergreen forests:

Kayea assamica	68.08
Dysoxylum sp.	6.35
Echinocarpus assamicus	3.70
Canarium sp.	4.31
Pterocarpus lancaefolium	6.25
Mesua ferrea	2.69
Terminalia chebula	2.57
Amoora wallichii	1.50

c) Nowgoang area: Stems per hectare are:-

Lagerstoremia speciosa-4.6, Bombax ceiba-2.1, Albizzaia procera-3.0, Castanopsis sp.-4.6, Syzygium praecox-1.6, Vitex peduncularis- 1.6,Terminalia belerica-2.6, Lannea coromendelia-2.0, Shorea robusta-5.2, Cassia fistula-6.0, Lagerstroemia speciosa-8.0,E. Parviflora-4.9, Bombx ceiba-2.1, Albizzia procera-2.2, Mansonia dipikae-1.6 Pterospermum acerifolium-1.9, Toona ciliata-2.2, Ulmus lancaefolia 4.0

Nagaland State

North Nagaland:

The forests are of tropical Evergreen and Semi-evergreen with many sub-types, juli forests have patches of bamboo and reeds.

Types of Forests	No. of stems per ha.
Evergreen and Semi-evergreen forests	205
Pure Bamboo shrubs:	69
Bamboo and Miscellaneous shrub	85
Reed strata	44

Principal Species with Stems per hectare:

Dipterocarpus macrocarpus-10.54, Lagerstroemia speciosa-0.09, Gmelina arborea-0.53, Dysoxylum procerum-4.44, Tetrameles nudiflora-0.69, Terminalia belerica - 1.75, Pterosperum lancaefolium - 6.14, Canarium resinifolium - 6.24, Castanopsis indica - 6.58, Terminalia myriocarpa- 1.17, Syzygium cumini- 2.00, Vatica lancaefolia -5.17, Endospermum chinensis- 2.29, Magnolia/ Michelia sp.- 2.29, Schima wallichii -2.44, Phoebe goalparensis- 0.14, Shorea assamica- 6.68, Mesua ferrea- 6.49, Artocarpus chaplasha 0.68, Bischofia javanica- 2.78, Michelia champaca - 1.70.

Assam State

North Cachar (Cachar hills):

Stems per hectare: (Less common trees):

Amoora wallichii	0.1	*Schima wallichii*	6.2
Lagerstoemia parviflora	3.6	*Phoebe cooperiana*	0.0
Phoebe goalparensis	0.5	*Mesua ferrea*	0.1
Ailanthus grandis	0.3	*Dillenia indica*	5.9
Tetrameles nudiflora	1.1	*Kydia calycina*	2.9
Chukrassia velutina	0.5	*Garuga pinnata*	2.1
Morus lavaegata	0.3	*Artocarpus sp.*	1.6
Terminalia sp.	8.2	*Boxbax ceiba*	0.4
Michelia sp.	0.3	*Artocarpus lakoocha*	0.8
Gmelina arborea	2.1	*Adina oligocephana*	1.9
Cinnamomum sp.	0.1	*Albizzia odoratissima*	0.2
Terminalia myriocarpa	0.0	*Bauhinia sp.*	12.2
Castanopsis indica	0.9	*Carega arborea*	5.7
Pterospermum	0.5	*Ficus sp.*	1.5

Meghalaya State

This survey pertains to a very small area in the foot hills of Southern slopes where shorea robusta occurs. Meghalaya has varied species occurring in Sub-tropical and Sub-temperate climate. Most of the species of medicinal value.

Trees Per hectare

Tectona grandis	135.22	Terminalia tomentosa	5.12
Ougenla oojeinensis	13.67	Anogeissus latifolia	12.72
Lagerstroemia parviflora	11.31	Boswellia serrata	3.52
Lananea coromendelica	5.36	Madhuca latifolia	1.41
Adina cordifolia	2	Mitragyna parviflora	2.24
Wrightia tinctoria	13.36	Grewia tiliaefolia	7.98
Other species	62.68	Shorea robusta	33.90
Bauhinia sp.	12.28	Terminalia sp.	8.22
Callicarpa sp.	7.08	Dillenia indica	5.97

Schima wallichi	6.22	albizzia sp.	5.83
Careya arborea	5.75	Others	87.75

In addition following species have also been recorded:

Broad-leaved stratum-168 economically important species have been identified - (229 trees per hectare). Occurrence of species per hectare are furnished below:

Castanopsis indica/Castanopsis sp. -14.10; altingia excelsa-7.23; ficus sp.-6. 56; quercus sp.-6.46; sygygium sp.-6.361; alnus nepalensis-4.66; Canarium sp. -4.66; Macaranga indica-4.41; Dysoxylum binecteniferum-4.10; amoora sp. -3.33; Calicarpa arborea-3.33; Terminalia myriocarpa- 3.28; Albizzia sp.-2.719; Canarium resiniferum- 2.61; Magnolia sp.-2.46; Trema orientalis -2.36; engelhardtia spicata-2.15; Mesua ferrea-2.15; Michelia sp. -4.92

Bamboo Stratum- 9 species are prominent *e.g.,*- Castanopsis, Altingia, Bischofia, Macaranga, Sterculia, Quercus Alstonia, Michelia, Schima (per hectare

Altogether 168 species of trees have been recorded alongwith their occurrence per hectare. Of these species most widely occurring trees have been depicted. The occurrence of Ficus, Gmelina, glochidion, Artocarpus, Antidesma, Cordia, dillenia, dysoxylum ,Erythrina, Eugenia, Terminalia species and species of several other tress indicate a bioecological bonanza for bird and animal fauna to survive.

Arunachal State

Lohit and Tirap Forests of Arunachal Pradesh:

Principal species of economic importance are:

Mesua ferrea, Dipterocarpus macrocarpus, shorea assamica, Michelia champaca, amoora wallichii, Mansonia dipikae, Canariun resinifolium, Kydia calycina, Pterospermum acerfolium, Bischofia javanica, Schima wallichii, ailanthus garndis, Gmelina arborea, Treminalia citrina.

Stems per hectare:

Dipterocarpus sp.	Shorea forests	226 stems/ha.
	In mixed miscellaneous Forests	150 stems/ha.

In Lohit forest	215 stems/ha.
In Changlang	185 stems/ha.
Khonsa	294 stems/ha.

So far volume per hectare are concerned, Terminalia myriocarpa Castanopsis sp., dillenia indica, toona ciliata, cinnamomum cecedodaphne and chukrasia tabularis show prominence, besides that of most other principal economic species.

West Bengal State

Kalimpong Hills of West Bengal:

According to a token survey carrid out over small area the following species were listed:

Machilus gammieana	22 stems/ha.
Lithocarpus lineata	15 stems/ha.
Alchimundra cathcartii	15 stems/ha.
Lithocarpus pachyphylla	10 stems/ha.
Nyssa javonica	9 stems/ha.
Castanopsis sp.	8 stems.ha.
Acer campbellii	8 stems/ha.
Cinnamomum sp.	8 stems/ha.

Meghalaya State

This survey pertains to a very small area in the foot hills of Southern slopes where shorea robusta occurs. Meghalaya has varied species occurring in sub-tropical and sub- temperate climate. Most of the species of medicinal value:

Trees per hectare	
Tectona grandis	135.2293
Terminalia tomentosa	5.1211
Ougenia oojeinensis	13.6774
Anogeissus latifolia	12.7263
Lagerstroemia parviflora	11.3199
Boswellia serrata	3.5248
Lannea coromendelica	5.3636
Madhuca latifolia	1.4197
Adina cordifolia	2.9277

Mitragyna parviflora	2.2456
Grewia tiliaefolia	7.9847
Wrightia tinctoria	13.3636
Other species	62.6827

In addition following species have also been recorded:

Shorea robusta	33:90
Bauhinia sp.	12.28
Terminalia sp.	8.22
Callicarpa sp.	7.08
Dillenia indica	5.97
Schima wallichi	6.22
Albizzia sp.	5.83
Careya arborea	5.75
Others	87.75

Trees presently not under threat:

Acacia arabica (Babul)
Alstonia scholaris (Chatni)
Anona squamosa (Atal)
Azadirachta indica (Neem)
Artocarpus heterophylla (Kanthal)
Bauhinia racemosa (Seta kanchan)
Bauhinia purpurea (Deba Kanchan)
Butea monosperma (Palash)
Cassia fistula (amaltus)
Dipterocarpus sp. (Garjan)
Dillenia pentagyna (Tantari)
Erythrina indica (Madar)
Fixus bengalensis (Bat)
Ficus glomerata (Jagna dumur)
Ficus cunea (Jaya dumur)
Holarhena antidysenterica (Kurchi)
Nyctanthes arbor-tistris (Siwli)
Stereospermum personatum (Parari)
Sterblus asper (Seora)
Ziziphus sp. (Kul)
Thespesia populnea (Pipli)
Toona ciliata (Toon)
Tamarindus indica (Tentul)

Aegle marmelos (Boel)
Albizzia lebbek (Sirish)
Areca catechu (Supari)
Anthocephalus chinensis (Kadam)
Anona reticulata (Nona)
Bauhinia variegata (Rakta kanchan)
Bombax ceiba (Semul)
Carica papaya (Papaya)
Dalbergia sissoo (Sisam)
Delonix regia (Krishnachura)
Eucalyptus tereticornis (Blue gum)
Emblica officinalis (Amloni)
Ficus religiosa (Aswathwa)
Ficus hispida (Kak dumur)
Grewia sp. (Phalsa)
Moringa olifera (Sajina)
Nerium odorum (Rakta Karabi)
Vitex negundo (Nishinda)
Trema orientalis (Kuail)
Tamarix sp. (Barha)
Tectona grandis (Teak)
Termarindus indica (Arjuna)

Syzygium cuminii (Kalo jam)
Prosopis cinararia
Diospyros melanoxylon (Kend)
Prosopis spicigera
Putranjiva roxbughii (Putranjiva)

Trees occurring sporadically which may be vulnerable unless felling restricted:

Acacia Suma (Somi)
Ailanthus excelsa (Mahanimba)
Anogeissus latifolia (Dhaw)
Boswellia serrata (Salai)
Citrus sp.
Gmelina arborea (Gamar)
Flacourtia sepiaria
Hydnocarpus sp. (Chalmugra)
Michelia Montana (Champ)
Mahonia rnepalensis
Ougenia oojeinensis (anjan)
Phoenix sp. (Khejur)
Pterygota alata (Narikeli)
Spondius mangifera
Knema augustifolia
Mellotus philippinensis (Rohini)
Polylthia semiarum (Labsi)
Aglaia roxburghii
Mangium salvaefolium
Barringtonia acutaugula (Hizal)
Lannea coromemdelica (Jeol)
Trewia nudiflora (Pitali)
Ochrocarpus longifolius
Disoxylum hamilitonii (Gab)
Evodia Lunnuankenda
Diospyros embriopteris (Gab)
Michelia champaca (Champ)
Boswellia serrata (Salai)
Capparis deciduas
Citrus sp.
Careya arborea (Kumbi)
Cochlospermum gossypium (Galgali)
Gmelina arborea (Gamar)
Acacia catechu (Khair)
Anacardium occidentale (Kaju)
Bassia latifolia (Mahua)
Capparis deciduas
Careya arborea (Kumbi)
Feronia elephantum
Hopea odorata
Melia azadirach (Goraneem)
Michlia niligerica (Champ)
Oroxylum indicum (Sona)
Premna sp. (Ginari)
Pongamia pinnata (Karanj)
Terminalia chebula (Haritaki)
Salix tetrasperma
Kigelia pinnata
Morus alba (Tut)
Bischofia javanica (Kanjal)
Adansonia digitata (Kalpataru)
Artocarpus lakoocha (Dewa)
Crataeva nurvula (Barin)
Symplolcos racemosa (Lodhra)
Flacourtia catafracta (Baichi)
Garuga pinnata (Giga)
Cordia dichotoma (Bahanari)
Evodia fraxinifolia
Mesua ferrea (Nageswar)
Antiaris toxicaria
Aglaia roxburghii (Priangn)
Adarnsonia digitata (Kalpataru)
Mangium salvaefolium
Artocarpus lakoocha (Dewa)
Barringtonia acutaugula (Hizal)
Crataeva nurvula (Barin)

Feronia elephantum
Flacourtia sepicaria
Hopea odorata
Hydnocarpus sp. (Chalmugra)
Melia azadirach (Goraneem)
Michelia montana (Champ)
Michelia niligerica (Champ)
Mahonia nepalensis
Ougenia oojeinensis (Panjan)
Premna sp. (Ginari)
Phoenix sp. (Khejur)
Pongamia pinnata (Karanj)
Pterospermum acerifolium (Kanak champa)
Pterygota alata (narikeli)
Terminalia chebula (Haritaki)
Spondius mangifera
Salix tetasperma
Knema augustifolia
Keigelia pinnata
Mallotus philippinensis (Rohini)
Morus alba (Tut)
Polylthia semiarum (Labsi)
Bischofia javanica (Kanjal)
Elaeocarpus ganitrus (Rudrakshya)
Pterocarpus marsupium (Peasal)
Strichnos potatorum (Nirmala)
Cochlospermum gossypium (Galgali)
Pterospermum acerifolium (Kanak champa)
Gardenia gummifera (Blurur, Neriherigar)
Gynocordia odorata (Bandre, Chakmugra)
Aphanamyxis polystachia (Tikiara)
Elaeocarpus tuberculatus (Rudraksha)

Lannea coromemdelica (Jeol)
Gardenia gummifera (Blurur, Neriheriger)
Sympolcos racemosa (Lodhra)
Trewia nudiflora (Pitali)
Gynocordia odorata (Bandre, Chalmugra)
Flacourtia catafracta (Baichi)
Ochrocarpus longifolius
Garuga pinnata (Giga)
Aphanamyxis polystachya (Tikiaraj)
Disoxylum hamiltonii (Gab)
Cordia dichotoma (Bahanari)
Elaeocarpus tuberculatus (Rudraksha)
Evodia lunuankenda (Rudrakshya)
Evodia fraxinifolia
Diospyros embriopteris (Gab)
Mesua ferrea (Nageswar)
Michelia champaca (Champ)
Antiarix toxicaria
Elaeocarpus ganitr
Juniperus sp.
Pterocarpus marsupium (Peasal)
Strichnos nux-vomica (Kuchila)
Juniperus sp.
Strichnos nux-vomica (Kuchila)

Trees (need immediate protection - a few selected species):

Balasamodendron mukul
Camphora officinarum
Caryota urens
Cerbera manghas
Cupressus sempervirens
Aquillaria agallocha (Aguru)
Taxus baccata
Schrebera switeniodes (Ghanta parul)
Bixa orillana
Callophyllum inophyllum
Ceiba pentandra
Juglans regia
Betula utilis
Commifera mukul (Guggal)
Juniperus communis

Some sporadically / occurring trees need protection :

Acacia suma (Som)
Acacia catechu (Khair)
Ailanthus excelsa (Mahanimba)
Anacardium occidentale (Kaju)
Anogeissus latifolia (Dhaw)
Bassia latifolia (Mahua)

Chapter Nine

Status Assessment of Shrub Flora

The author has drawn a broad survey assessment of the shrub vegetation of the country having medicinal properities.

The legend drawn shows the Status of each species. It may be seen that maximum number of species has been marked as 'S' *i.e.*, of Sporadic occurrence. Sporadic occurrence may be in 'Single' or in 'groups' of several individuals. The second group *i.e.*, groups of several individuals have been marked with asterisk as S*. But it is to realize that this assessment is made on a broad basis as authentic figure will emerge only after a statistically sound plant analysis design and field study.

The list is comprehensive and gives a view of the common and important shrubly medicinal plants of India. So it may be easy to draw a conservation strategy.

Results of Study Over the Country

The distribution of few shrubs as a result of analysis shows the follwing:

I. Few shrubs have wide distribution.

II. Several shrubs are conspicuous all over the country.

III. Few shrubs are gregarious in their occurrence.

IV. Most of the shrubs are confined to forested regions.

V. Moist localities have more of shrubs than other habitats.

VI. Ground cover of shrub layer has many tree species as regeneration.

The degree and duration of the shade have a positive bearing on the composition and density of shrub crop. Heavy shade results in a very limited number of shrub flora. The intensity of crop is concerned with photosynthesis and food production, it also effects flower and fruit production. Specific insect and avaifaunal species are concerned with shrubs for food and pollination. At least eight shrub genera form an imperative source of larvae food of the moths and butterflies. Over-exploitation of tree crops in all the three storeys in the forests cause considerable physical damage to the shrubs; these shrubs are essential for soil binding and create a micro climate for the survival of young plant regeneration; besides, they help litter formation and conservation of water.

Shrubs and young regeneration of other plant species suffer irreparable losses due to repaeated grazing and fire. These two factors are directly responsible for eradication of plant species where only those with underground stem survive. The author's research reveal that only about 40 to 50 per cent of the species survive such depredation; also the density of occurrence of species is affected. Vast tracts of forest where grazing and fire occurs have now very coarse grass and a few are hardy shrub species.

About ninety species of shrubs are conspicuous in the temperate hills of India some of which have colourful flowers; some also grow in profusion. Special mention may be made of rhododendron and roses. In the plains, however, more than one hundred conspicuous shrub species occur but they do not bean conspicuous flowers. In eastern india there are at least a hundred species of fodder shrubs; it is about two thirds in other parts of the country. There are at least forty genera of shrubs yielding edible fruits which attract avifauna, animals and men alike.

A survey made by the author all over the country has revealed that only fifty species of shrubs are commonly found. Lantana, Eupatorium, Clerodendrum, Calotropis, Cassia, Carissa, Capparis, etc, are found conspicuous and are occurring in profusion all over the country. It should be considered as an ominous indication of systematic disappearance of species primarily due to anthropogenic factors. The number of shrub species in the country is about 4500 (tree species about 1200); but wide occurrence of only 50 species is hardly to be believed which is too meagre.

Status of Medicinal Shrubs

Vast shrub resources of this country have not been studied from medicinal, aesthetic and environmental points of view. This country has shrub flora occurring on wide range of alkaline, saline, esturine sandy soils in varied climatic conditions. The time is ripe to select typical shrubs that have conspicuous flowers wide range of foliage patterns and shapes and plant them in arboreta, parks and gardens. Some shrubs are indispensibte ingredients of parks, gardens, residential quarters premises and in environmental planting to arrest air pollution.

An attempt has been made to list some common and conspicuous shrubs occurring all over the country (Statewise) and to mark some common medicinal plants. Although some of these shrubs are obnoxious weeds (Eupatorium, Cleistanthus, Lantana, Calotropis, Euphorbia, Jatropha etc. Quite a good number of these shrabs occur on dry rocky and sandy areas (Euphorbia, Opuntia, Agave, etc.) which provide shade, protect soil, shelter birds and bear fruits for man and animal besides, providing plants of medicinal values. Vast avifaunal species in the country side also derive a lot of food and seek shelter in such shrub flora.

India has about 265 species of woody (Lianas) and herbaceous climbers; of these about 125 are woody climbers and the rest are herbaceous.

Of these, a hundred species are used as medicinal plants. Climbers which are trailors and the herbaceous ones which depend fully on the host plants for support have no bright future as their existence depend normally on host plants.

Some woody climbers such as Tinospora, Pureria, Entada, Vitis, Mucuna etc. require strong host plants for their support and survival and therefore, removal of associate trees bring their obvious depletion.

Quite a large number of herbaceous climbers grow in forest margins, waste places and on marginal lands; they very frequently face biotic hazards and therefore, their existence as a result become increasingly threatened; the biotic pressure is in the form of heavy grazing by animals growth of human population and their dependence on forests and demand of land for settlements.

A list of commonly available climbers are presented with comments on their status:

Over-exploited-may face depletion:

- Paederia foetida (Gandhal) -Very sporadic:

- Tinospora cordifolia (Gulancha): Over-exploited by forest dwellers, it being a good fodder species (enhance milk formation in cow and also for collection for medicinal purposes. It is typically a forest species. Other plants are:

- *Cuculus villosa* (jaligamni)
- *Stephania harnandifolia* (Nimutha)
- *Cissampelos pareira*
- *Cyclea burmani*
- *Vallaris solane* (Haparmati)
- *Tylophora indica* (Antomal)
- *Gymnema sylvestre* (Chlagalbeti)
- *Jasminum multifloram* (Kanda)
- *Aganosma dichotoma* (Madhabilata)
- *Tinospora malabarica*
- *Calamus viminalis* (Bet)

(These plants grow in forest margin, as such they are directly or indirectly destroyed due to lopping and pilferage of host trees.).

- *Asparagus Sp.*
- *Dioscorea sp.*
- *Tribulus terrestris*
- *Gnetum sp.*
- *Aristolochia indica*
- *Smilax macrophylla*

Shrub Flora (With association of herbs) of Waste Places (Sporadic):

Table 9.1

Species	Status	Species	Status
Sida vernicoefolia (joka)	S	*Sida rohmboidea* (Pila Barela)	S
Sida cordifolia (Barlela)	S	*Urena lobata* (Ban Okra)	S
Urena sinuala	S	*Clitoria turnatea*	C
Acalypha indica (Muktajhuri)	S	*Centella asiatica*	S
Evolovulus asinoides	S	*Merremia emerginata*	S
Solanum nigram	S	*Solanum ferox*	S
Solanum indicum	S	*Solanum xanthocarpum*	S
Solanum torvum	S	*Solanum surattense*	S
Physalis peruviana	S	*Withania somnifera*	C/D
Scoparia dulcis	S	*Picrorrhiza kuroa*	C/D
Chenopodium album	S	*Abutilon indicum* (Paleri)	S
Heliotropium indicum (Srihaslini)	S	*Flemingia* sp.	S
Zornia diphylla	S	*Trianthema portulacastrum*	S
Musaenda frondosa (Nagballi)	C	*Gangrea maderaspetana*, (Namuti)	S
Lorathus sp.	S	*Argemone mexicana* (Sealkanta)	S

contd.....

Table 9.1 contd...

Species	Status	Species	Status
Capsella bursapasteris	S	*Cleome gynandra*	S
Clemoe isosandra	S	*Polygala chinensis* (Muradu)	S
Polygala crotalarioides (Nilkanti)	S	*Polycarpea corymbosa*	S
Portulace oleraccea (Barnunia)	S	*Portulaca quadrifida* (Chota nunia)	S
Malvastrum sp.	S	*Torenia asiatica*	S
Vandellia sp.	S	*Justicia gandurussa* (Jagat madan)	S
Rungia parviflora (Piri)	S	*Achyranthus aspera* (Apang)	S
Pogostemon parviflorus	S	*Boerhaevia diffusa* (Punarnava)	S
Amaranthus spinosus (Kanta note)	S	*Amaranthus gangeticus* (note)	S
Amaranthus viridis	S	*Xanthium strumerium* (Ban Okra)	S
Canabis sativa	C	*Rumex dentatus*	S
Rumex maritimus	S	*Cyanotis sp.*	S
Phyllanthus reticulatus (Panjuli)	S	*Phyllanthus simplex*	S
Ipomoea fistilosa		*Calotropis giggantia* (Akanda)	S
Datura stamonium (Sadadhanara)	S	*Datura stramonium* (Sadadhanara)	S
Clerodendrum serratum	S	*Glycosmis arborea*	S
Martinia diandra (Baghnakhi)	S	*Grewia sp.* (Phalsa)	S
Moringa olifera (Sajina)	C	*Nyctanthes arbortistris* (Siwli)	C
Nerium odorum (Rakta karabi)	C	*Streblus asper* (Seora)	S
Vitex negundo (Nishinda)	S	*Ziziphus sp.* (Kul)	S
Trema orientalis (Kuail)	S	*Thespesia populnea* (Pipli)	C
Tamarix sp. (Barjha)	S	*Toona ciliata* (Toon)	C
Tectona grandis (Teak)	C/PS	*Tamarindus indica* (Tentul)	C
Termarindus indica (Arjuna)	C/PS	*Syzygium cuminii* (Kalo jam)	C
Prosopis spicigera	PS	*Prosopis cinararia*	S/PS
Putranjiva roxburghii (Putranjiva)	C/PS	*Diospyros melanoxylon* (Kend) (Parari)	S
Holarrhena antidysenterica (Kurchi)	PS	*Stereospermum personatum* (Parari)	PS
Phyllanthus urinaria (Hazarmani)	S	*Phyllanthus niruri*	S

contd.....

Table 9.1 contd...

Species	Status	Species	Status
Cenopodium ambrosioides	S	*Wedelia calendula* (Bhimraj)	S
Euphorbia hirta (Swadaparui)	S	*Malachra capitata*	S
Atylosia scarabiodes	S	*Sophora sp.*	S
Mallugo hirta	S	*Oldenlendia herbacea*	S
Elephantopus scaber (Shaydalan)	S	*Celosia argentea*	S
Polygonum orientale	S	*Oxalis corniculata* (Amrul)	S
Oxalis acetosella	S	*Martynia diandra* (Bagh nakhi)	S
Pandanus tectorius (Keya)	V	*Pandanus roxburgiana* (Keya)	D.
Breynia rhamnoides	S	*Breynia patens*	S
Datura stramonium (Sadadhatura)	C	*Ervatamia coronania* (Tagar)	
Japropha gossypifolia (Bag veranda)	S	*Thevetia nerifolia* (Halde karabi)	
Caesalpinia pulcherrima (Krishnachura)	C	*Hibiscus mutabilis* (Sthalpadma)	
Barleria cristata (Swata jhanti, Kalajhanti)	C	*Lawsonia inermis* (Mehendi)	
		Ricinus communis (Aronda)	
Sporadic (not threatened)			
Mimosa pudica (Lajjabati)	S	*Cassytha filiformis* (silinga)	S
Capparis spinosa	S	*Croton tiglium*	S
Mimosa rubicaulis (Dhuna)	S	*Ipomoea reptans* (Kalmi Sag)	S
Ageratum conyzoides	S	*Blumea lacera* (Kukhim)	S
Eclipta alba	S	*Taraxacum officinate*	S
Biophytum sensetivum (Ban naranga)	S	*Helectris isora* (Avikush, Athmum)	S
Sphaeranthus indica (Maharashrabani)	S	*Justicia adhatoda* (Vasak)	C
Chrosophora cincera (Tut kukusima)	S	*Morinda tinctoria* (Abecheek)	S
Lindenbergia urticaefolia (Haldebasanta)	S	*Coffea bengalensis* (Bancoffee)	S
Andrographis paniculata (Kalmegh)	D	*Murraya exotica* (Kamini)	C/S
Oldenlendia corymbosa (Khetkapra)	S	*Ixora coccinia* (Rangan)	C
Capparis decidua	P/S	*Butea crispa* (Banchalida)	C

contd.....

Table 9.1 contd...

Species	Status	Species	Status
Abrus precatorius (Kunch)	S	*Murrya koengii* (Curry leaf)	C/PS
Hiptage bengalensis (Madhabilata)	S	*Justicia adhatoda*(Vasak)	C/PS
Jasminum grandiflorum (Chameli)	C	*Morinda citrifolia* (Anch)	S
Argyreia speciosa (Rishya gandhri)	S	*Coffea sp.*	S
Marsdenia volbilis (Nach chikni)	S	*Ixora passiflora* (Gandhale, Rangan)	C
Hemidesmus indicus (Anatamul)	D	*Laportia crenulata* (Chutra)	S

Temperate areas (Threatened):

Species	Status	Species	Status
Delphinium denudatum (Nilbishi)	D	*Cimicifuga foetida(Jimti)*	D
Paeonia emodi(udsalap)	D	*Illicium griffithii*	D
Digitalis purpurea	C	*Swertia chlrata*	D
Geranium nepalense	D	*Solanum dalcamara*	D
Viscum sp.	D	*Fritillaria sp.*	D
Lilium gigatium	D	*Vivurnum foetidum*	S
Hypericum patulum	D	*Cinchona succirubra*	C
Lonicera glauca	D	*Osbeckia nepalensis*	D
Cinchona ledgerina	C		

Shrubs of Temperate Hills (under no threat)

Melastoma matabathricum	S	*Woofodia fruticosa (Dhaiphal)*	S
Berberis aristata (Daru haridra)	C	*Memecylon edule*	S
Pogostemon plectranthoides	S	*Pogostemon parviflorus*	S
Colebrookia oppositifolia	S	*Anisomeles indica*	S
Hypericum patulum	S	*Hypericum sp.*	D
Vivenum foetidum	S	*Lonicera glauca*	S
Cinchona succirubra	C	*Cinchona ledgerina*	C
Osbeckia nepalensis	S	*Osbeckia crinita*	S

Stray shrubs in tropics under no threat:

Cassla tora (Chakmudl)	S/PS	*Cassia sophera*	S
Cassia occidentalis (Kal kasundi)	S/PS	*Euphorbia nerifolia* (Snowii)	S

contd.....

Table 9.1 contd...

Species	Status	Species	Status
Gossipium herbaceum	C	*Eupatorium odoratum*	S
Lantana camara	PS	*Leonurus sibiricus*	S
Hiptis suaveolens	S/PS	*Boerhoevia diffusa*	S
Boerhaavia repens	S	*Artemisia vulgaris*	D
Curcuma longa	C	*Indigofera pulchella*	S
Indigofera linifolia	S	*Indigofera trifolia*	S
Indigofera sp.	S	*Demodium gangeticum*	S
Desmodium polycarpum	S	*Desmodium pulchellum*	S
Desmodium sp.	S	*Crotalaria verrucosum*	S
Crotalaria juncea	C	*Crotalaria albida*	S
Crotalaria sp.	C	*Alysicarpus longifolius*	S
Uraria picta (Sankajota)	S	*Urariaogopoides* (Chakulia)	S
Psoralia corylifolia (Buchki)	S	*Costus speciosus*	D
Alpinia glalanga	S	*Alpinia allughus*	S
Curculigo orchiuoides	S	*Tacca pinnatifida*	D
Tragia involucrate	S	*Pouzolzia indica*	S
Sanseveria roxburghiana	C	*Curcuma amgustifolia*	S
Presently safe species, not threatened (Tropics)			
Mimosa Pudica (Lajjabati)	S	Ervatamia coronaria (Tagar)	C
Cassytha filiformis (Silinga)	S	Thevetia nerifolia (Halde karabi)	C
Capparis spinosa	S	Hibiscus mutabilis (Sthal padma)	C
Croton tiglium	S	Hibiscus albomoschus (Latkasturi)	C
Mimosa rubicaulis (Dhuna)	S	Ricinus communis (Aronda)	C
Jasminum gradiflorum (Chameli)	C	Calotropis gigantia (Akanda)	S
Argyreia speciosa (Rishya gandhri)	S	Datura fastuosa (Dhatura)	S
Ipomoea reptans (Kalmi Sag)	C	Nerium odorum (Rakta karabi)	C
Capparis decidua	S	Hibiscus cosa-sinnensis(Jaba)	C
Arbus precatorius (Kunch)	S	Lawsonia inermis (Mehendi)	S
Hiptage bengalensis (Madhabilata)	S	Punica grnatum (Darimba)	C
Datura metal (Dhatura)	S		

contd.....

Table 9.1 contd...

Species	Status	Species	Status

Cucurbetaceous Climbing Shrubs:

e.g., Cucumis, Momordica, Luffa, Lageneria, Trichosanthes and Beniscasa, others are:

Marsdenia volubilis (Nach chikni)

Hemidesmus indicus (Anantamul)

Croton caudatus

Stray Shrubs under no Threat:

Species	Status	Species	Status
Cassia tora (Chakmundi)	S	*Woodfordia fruticosa* (Dhaiphal)	S
Cassia occidentalis (Kal kasundi)	S	*Memecylon edule*	S
Gossipium herbaceum	S	*Pogostemon parviflorus*	S
Lantana camara	PS	*Anisomeles indica*	S
Hiptis suaveoens	S	*Neraim odorum* (Rakta Karabi)	
		Hibiscus rosa-sinensis (Jaba)	C
Cassia sophera	S	*Hibiscus albomoschus* (Latkasturi)	C
Euphorbia nerifolia (Snowii)	C	*Punica granatum* (Darimba)	C
Eupatorium odoratum	PS	*Datura metal (Dhatura)*	S
Leonurus sibiricus	PS	*Calotropis procera* (Akanda)	
Boerhaavia diffusa.	S	*Datura fastuosa* (Dhatura)	S
Boerhaaria repens	S	*Jatropha gossypifolia* (Bag venda)	S
Clerodendrum serratum	S	*Barleria cristata* (Swata jhanti, Kalajhanti)	C
Martinia diandra (Baghnakhi)	S		
Artemisia vulgaris	D	*Pandanus roxburghiana* (Keyal)	S
Glycosmis arborea.	S	*Breynia patens*	S

Special Groups of Plants:

– *Aromatic plants:*

Mimusops elengii (Bakul)	Common
Vetiveria zizanioides (Vetiver)	Common
Jasminum grandiflorum (Jasmine)	Sporadic
Pandanus odoratissima (Screw-pine)	Sporadic
Myristica fragrans (Nulmig)	Rare

contd.....

Table 9.1 contd...

Species	Status	Species	Status
Syzygium aromaticum (Clove)		Rare	
Cinnamomum camphora (Camphor)		Rare	
Aquilaria agallocha (Aloewood)		Rare	
Santalum album (Sandle wood)		Very localized (Protected)	
– *Cosmetic plants (Mostly cultivated)*			
Lawsonia inermis (Henna)		C	
Sesamum indicum (Sesame)		C	
Hibicus rosa-sinensis (Hibiscus)		C	
Acacia catechu (Areca nut)		C	
Sapindus trifoliatus (Soap-nut)		C	
Curcuma longa (Turmetic)		C	
Mallotus philippinensis (Indian Kamella)		PS	
Citrus indica (Lime)		C	
Crocus sativus (Saffron)		C	
Piper betel (Supari)		C	
Indigofera tinctoria (True indigo)		S	
Hemidesmus indicus (Sarsaparila)		D	

– *Sacred Plants (common)*

Species	Status	Species	Status
Ficus bengalensis (Banyon)	PS	*Ocimum sanctum* (Holibasil)	C/Ps
Butea monosperma (Palash)	PS	*Cocos nucifera* (Coconut)	C/PS
Aegle marmelos (Boel)	PS	*Cannabis sativa* (Indian hemp)	C
Ficus religiosa (Sacred fig)	PS	*Syzygum cuminii* (jamun)	PS
Azadirachta indica (Margosa)	PS	**Nelumbium speciosum* (Lotus)	D
Elaeocarpus ganltrus (Rudrakshya)	D		
Catheranthus roseus (Nayantara)	C		
– *Culinary Plants (mostly cultivated)*			
Moringa olifera (Drum stick)	C	*Tamarindus indica* (Tamarind)	C
Allium sativum (Garlic)	C	*Zingiber officinale* (Ginger)	C
Mangifera indica (Mango)	C	*Cinnamomum tamala* (Tejpata)	C

contd.....

Table 9.1 contd...

Species	Status	Species	Status
Borassus flabellifer (Palmyra)	C	*Bassia latifolia* (Mahua)	PS
Nigella sativa (Kalijira)	C	*Cuminum cyminun* (Cumin)	C
Solanum melonginum (Egg plant)	C	*Carum acpticum* (Joan)	C
Carum roxburgiana (Randhuni)	C	*Foeniculum capillaceum* (Mouri)	C
Ferula asafetida (Hing)	C	*Piper nigrum* (Black piper)	C
Cinnamomum zeylanicum (Chinamon)	C		
Coriandrum sativum (Corriander)	C		
Elettaria cardamomum (Cardamom)	C		
– *Other Plants of Tropics Temperate Areas :*			
Zingiber officinale	C	*Zingiber cassumaner*	S
Commelina obiqua	S	*Pathos scandens*	C
Scindapsus officinalis (Rajpipal)	C	**Amomum subulatum*	C
**Elleterla cardamom*	C	**Rauwolfia serpentina*	D
Tephrosia purpurea	S	*Tephrosia villosa*	S
**Curcuma Zedoraria* (Kkangi)	C	**Curcuma amada*	C
**Curcuma aromatica* (Halud)	C	**Cannabis sativa*	C
Polygonum plebejum	S	*Polygonum glabrum*	S
Polygonum barbatum	S	*Polygonum hydropiper*	S
Polygonum orientale	S		

Distribution zones of Conspicuous shrub Flora in India:

Madhya Pradesh:

Holarrhena antidysenerica	*Acacia catechu*
Butea monosperma	*Lantana monosperma*
Phoenix sylvestris	*Osbeckia aspericaulis*
Helicteres isora	*Barleria prionitis*
Dichrostachys cinerea	*Zizyphus sp.*
Vitex negundo	

contd.....

Table 9.1 contd...

Species	Status	Species	Status
Maharashtra:			
Calotropis procera		*Cassia auriculata*	
Indigofera gerardiana		*Zizyphus mauritiana*	
Lantana camara		*Carissa carandas*	
Zizyphus sp.		*Cassia sp.*	
Plectranthus rugosus		*Cleistanthus collinus*	
Gymnosporia spinosa.			
Manipur:			
Desmodium cephalotes		*Pueraria hirsute*	
Priotropis cytisoides		*Buddleia sp.*	
Cycas siamensis		*Phoenix humilis.*	
Orissa:			
Barleria prionitis		*Zizyphus sp.*	
Calotropis procera		*Adhaoda vasica*	
Anona squamosa		*Cassia sp.*	
Combretum decandrum		*Zizyphus mauritiana*	
Jatropha glandulifera		*Vitex negundo*	
Punjab:			
Capparis decidua		*Tamarix dioica*	
Zizyphus nummularia		*Prosopis juliflora*	
Salvadora oleoides		*Calotropis proceram*	
Butea monospera		*Adhatoda vasica*	
Murraya koenigii		*Lantana camara*	
Acacia arabica		*Carissa opaca*	
Capparis sepiaria		*Dodonaea viscose*	
Rajasthan:			
Acacia arabica		*Prospis juliflora*	
Zizyphus nummularia		*Cassia mimosoides*	
Calotropis procera		*Capparis deciduas*	
Mimosa rubicaulis		*Butea monospera*	
Acacia catechu		*Euphorbia nivulla*	

contd.....

Table 9.1 contd...

Species	Status	Species	Status
Lantana camara		*Prosopis spicigera*	
Acacia leucophloea		*Phoenix sylvestris*	
Solanum seaforthianum			
Tamil Nadu:			
Ixora parviflora		*Suaeda sp.*	
Zizyphus nummularia		*Cassia auriculata*	
Carissa carandas		*Dodonaea viscosa*	
Gardenia gummifera		*Cassia siamea*	
Acacia leucophloea		*Calotropis procera*	
Zizyphus sp.		*Opunita sp.*	
Randia dumetorum		*Hyupericum mysoremsis*	
Lantana camara		*Phoenix sylvestris*	
Rubus sp.			
Uttar Pradesh:			
Adhatoda vasica		*Zizyphus sp.*	
Moghania sp.		*Carissa sp.*	
Butea monsperma		*Zizyphus nummularia*	
Lantana camara		*Phoenix decidua*	
Scindapsus officinalis (Gajapepul)			
Dodonaea viscosa			
Jatropha gossypifolia			
Morinda tinctoria			
Randia dometorum			
Assam:			
Eupatorium sp.		*Lantana camara*	
Phoenix humilis		*Litsaea umbrosa*	
Rubus ellipticus		*Osbeckia sp.*	
Randia dumetorum			
Bihar:			
Phoenix sylvestris		*Butea monosperma*	
Carissa carandas		*Carissa opaca*	

contd.....

Table 9.1 contd...

Species	Status	Species	Status
Delhi:			
Calotropis procera		*Capparis deciduos*	
Cassia mimosoides		*Zizyphus mauritiana*	
Z. nummularia		*Prosopis species*	
Gujarat:			
Calotropis procera		*Zizyphus mauritiana*	
Capparis deciduas		*Cassia auriculata*	
Butea monosperma		*Cassia mimosoidessopis*	
Mimosa rubicaulis		*Acacia arabica*	
A. leucophloea		*Euphorbia royleana*	
Zizyphus nummularia		*Prosopis juliflora*	
Acacia catechu		*Helicteres isora*	
Suaeda fruticosa		*Premna coriacea*	
Himachal Pradesh:			
Acacia caesia		*Berberis sp.*	
Carissa spinarum		*Flacourtia indica*	
Berberis aristata		*Desmodium gangeticum*	
Prinsepia utilis		*Colebrookea oppositifolia*	
Indigofera geradiana		*Woodfordia fruticosa*	
Cotoneaster microphylla		*Plectranthus rugosus*	
Adhatoda vasica		*Berberis chitria*	
Dodonaea viscose		*Agave Americana*	
Rhus continues		*Zanthoxylum alatum.*	
Jammu & Kashmir:			
Calotropis procera		*Adhatoda vasica*	
Acacia modesta		*Clerodendrum viscosum*	
Lantana camara		*Viburnum foetens*	
Arisaema helleborifolium		*Desmodium tiliaefolium*	
Deutzia corymbosa		*Jasminum humile*	
Rubus lasiocarpus		*Berberis lycium*	
Robinia pseudo-acacia		*Pistacia integerrima*	

contd.....

Table 9.1 contd...

Species	Status	Species	Status
Rosa moschata		*Rubus ellipticus*	
Plectranthus rugosus		*Artemisia vulgaris*	
Indigofera pulchella		*Daphne sp.*	
Parrotia sp.			
Kerala:			
Cassia mimosoides		*Euphorbia tirucalli*	
Randia dumetorum		*Cocos nucifera*	
Pandanus sp.		*Phoenizx sylvestris*	
Lantana camara			
Karnataka:			
Acacia leucophloea		*Butea monosperma*	
Azadirachta indica		*Dichrostachys cinerea*	
Lantana camara		*Rhus wallichii*	
Calotropis procera		*Zizyphus sp.*	

Chapter Ten
Status Assessment of Herb Flora

The author has drawn a broad survey assessment of the herb vegetation of the country having medicinal properties.

The legend drawn shows the status of each species. If may be seen that maximum number of species has been marked as 'S' *i.e.*, of Sporadic occurrence. Sporadic occurrence may be in 'Single' or in 'groups' of several individuals. The second group *i.e.*, groups of several individuals have been marked with astarisk as S*. But it is to realize that this assessment is made on a broad basis as authentic figure will emerge only after a statistically sound plant analysis design and field study.

The list is comprehensive and gives a view of the common are important shrubly medicinal plants of India. So it may be easy to draw a conservation strategy.

Status of Herbs

The word "Weed" (meaning troublesome, undesirable and agggressive plants) is a misnomer to an ecologist or a conservationist. Human knowledge and ingenity may not reach perfection to unearth various properties and utility properties in them, yet they remain resourceful. A quantitative assessment of the flora in the region presents a distressing feature of the herb Flora of the region. Relentless biotic pressure from men for extension of cultivation, fire, grazing and other natural factors as flood, erosion etc. have, (i) eradicated many species from their original home, (ii) affected the diversity of genera and species, (iii) ousted soft species in favour of coase and resistant species particularly of few selected species of herbs and mainly shrubs, (iv) diluted the species both in number of species can be found wide spread and regenerating, while innumerable species recorded in old flora are hardly found growing in profusion.

Quantitative assessment of herb flora has revealed reduction of at least 50 per cent of the species from their natural home and the existing ones have been very much diluted and some receded. Over exploitation of all species of *Orchids, nymphaea, nelumbium, Primula, Androsace, Areneria, Meconopsis, Rheum, Coptis, Aconitum, Nardostachys, Anaphalis, Swertia, Selaginella* etc. and other medicinal and decorative species are also responsible for ruthless eradication of such species.

Shifting cultivation for centuaries have reduced vast forested tracts into grassland or Shrub land and reduction in number of genera and species. Even the plants recorded by Sri Hooker in mid ninetieth century (130 years ago) are not to be seen in those areas where biotic interference has been a regular feature.

The feature of landscape is fast changing. Shifting cultivation continues all over eastern India from foot hills to an altitude of 2000 m. As such flora is sub-tropical, lower temperate hills has been getting persistant blow; besides, the cultivation of Rubber, Coffee, Cocoa, Tea, Teak and Eucalyptus and other species as pure crop (Monoculture) over assessment would only reveal correct picture. Several orchids, ferns, fern allies medicinal and flowering plants of this region have already been recorded as rare and endangered and a bleak future is indicated.

Moist situation in the foot hills and temperate hills, account for richness of species. The ecotone regions in the foot hills and middle hills are obviously the richest zone of various flora and faunal crop and deserve immediate preservation. Diversity of plants will mean diversity of insects birds and animals.

Many shrub and herbs have been exploited heavily for fodder, besides grazing animals (Wild Life and domesticated animals) take a heavy toll of such vegetation. Most of the plant of Urticaceae form good fodder and are heavily exploited. Besides, many species of ferns, *Monocots, Flemengia, Desmodium, Saxifraga, Pouzolzia, Pilea, Elatostemma, Polygonum, Fagopyrum, Rumex, Limnophylla, Lippia, Phyla* are also grazed. Several monocot herbs (*Floscopa, Grasses, Sedges, Commelina, Cyanotis, Costos, Alpinia, Hypoxis, Cautleya* etc.) are heavily grazed for fodder.

Ecological studies on herbs cannot be done in isolation. Herbs shrubs and trees that regenerate along with grasses and sedges form an inseparable plant community. Each influences the other single or the groups. Tree species regeneration are mixed with herbs and shrubs at different stages of their growth and such regeneration forms about 50 per cent of the total species at any site either in shrub or herb layer. As such here-shrub associations shrubs-tree association or herb-tree associations are often conspicuous. More conspicuous of course are obnoxious shrub layers of *viz.*, Clerodendrum, many climbers forming tangle Capparis, Cassia, many species of Labiatae, Phlogacanthus, Woodfordia and very many other species. Such tough and resistant herb layers are hardly found among the shrubs. Some sembalance may be observed in thick mat forming borreria hispida (Rubiaceae) which cover extensive plantation areas in Terai and Duars of North Bengal and various eastern Indian states.

Study of insect and animal association, action may interaction in grassland or herbland is a separate subject and should be undertaken in full detail.

In the heavy shades of tropical and sub-tropical forests there is meager occurrence of dicot herbs. Comparatively however, there are abundance of grasses, sedges and some monocot herbs (Zingiber, Curcuma, Alpinia, Dioscorea, Commelina, Globba, Smilax etc.) On the whole, species of shrub-flora is richer in such tropical region that the herbs, while in temperate hills herbs-flora is richer, monocot, however, maintains same proportion.

Herbs are light demander. Alpine grass lands have vast number of herbaceous flora which display a vast array of colours.

Herbs render shelter to snakes lizards, frogs and many other ground dwelling animals. They afford protection to soil; in fact, they form the last line of defence against direct hit from sunrays and rains and protect soil from erosion and retention of moisture.

Status of Medicinal Herbs

Of all life forms of plants in India, the number of herb species outnumber others. Roughly, the number of shrub species is about double the number of tree species, while the number of herbs is about four times the shrubs.

Herb has been defined as "Plants with no woddy stems above the ground." Others define it as "A plant the stem of which dies every year" or as, "A plant with no persistent parts above the ground, as distinct from shrubs and trees". It has been defined in many other books ranking it with weeds which are not valued for beauty, that grow wild and hinder the growth of superior species, it has been found to grow where it is not desired and where it grows luxuriantly and plentifully. But little about their values as medicinal plant have been discussed.

No plants are weeds. Each species, has a part to play, either singly or as a member of the community. Each maintains a mutual harmony in it's niche and each niche plays a positive role in moulding theenvironment. Some berbs may be troublesome, most aggressive, harmful or annoying to man or to his agriculture and are termed as weeds. The herbaceous weeds grow luxuriantly; they are resistant, have a high reproductive capacity, gregarious, have a high level of seed dormancy, are quick growing often poisonous, deep rooted, annual or perennial and abundantly seed bearing. Several herbs are parasitic, some are hosts of fungal and baterial diseases, some are hosts of nematodes and some are associated with insects. Many are delicious food for animals and man alike; many are useful as medicinal plants. Some weeds are guardians of soil. They provide protective cover against erosive action of rain drops and run off. They are also used as mulch in cultivated land.

The herbs come in profusion in swamps, marshes and on other sites which accumulate water in rains. In permanent ponds, weeds also predominate. Aquatic life burst into activities and complete their life cycle within a short span. Succession of plants and animals goes on simultaneously on vertical and horizontal strata. Herbs (submerged, floating, emergent) growing in profusion have diversified animal periphyton of insects, amphipods, mites, snails, etc. The floral and faunal complex attract aquatic birds. The ponds, marshes and swamps become roosting sites and feeding ground for many species of birds. Heron feed on shallow water, grebe, cormorants and terns feed on fish in open water, the egrets, bitterns eat fish in shallow water, cranes and coots are omnivorous, rails, ibises, stilts and snipes probe around in mud, but various ducks eat seeds, roots and soft parts of weeds. In India the aquatic flora consist of 171 to 200 species of plants belonging to 90 genera and 39 families.

But of late the areas of water bodies have been found dwindling rapidly. Marshes and lowlying lands are being brought under human settlements.

Besides the aquatic sites, herbs also grow on road sides, rice field, dry cultivated land and in various types of forests. In spite of their wide range of occurrence relentless biotic pressure, flood and erosion have reduced the number of species and many have been threatened to the verge of extinction. Over exploitation of medicinal, decorative and other commercial species has been responsible for shrinkage of many species. Shifting cultivation has also led to the complete eradication of many species. Many herbs have also been over-exploited for fodder purpose.

The environmental role of trees, shrubs and herbs should be studied in conjunction with grasses, sedges, palms, epiphytes, parasites etc. as all these lifeforms constitute in separable components of the plant communities to mould the environment and enrich the medicinal plant resources.

Herbs form eighty per cent of the medicinal plant species. They need meticulous protection. There are many diffused and inconspicuous herbs which have showy flowers. Many herbs grow in grasslands, many in tall reeds and many in alpine grasslands. Innumerable herbs growing in temperate hills have also showy flowers. Many species of canes, yams, lily, asparagus, etc. which form a number of medicinal plants are being over-exploited. There are many species of aroids besides wild banana, ginger, turmeric, cardamom which grow over extensive areas and form conspicuous physiognomy of the forest floor. Such resources are being damaged by fire and grazing also.

The enumeration of the essential qualities of herb layer remains incomplete unless some mention are made of climbers, ferns and epiphytes.

Large woody climbers, another lifeform and a biotic association are remarkable because of their fantastic forms. They add complexities to the vertical structure of forests. There are, however, delicate climbers, shrubby climbers, woody climbers and stranglers apart from root climbers like some climbing palms. There are about six hundred climbers the half of which are shrubby climbers.

Epiphytes, however, are another lifeform that derive support from host plant. Some have remarkable adaptation to hold water. They consume water in a restricted way and many conserve water in their tubers. Base of overlapping leaves of some ferns forms a special niche where soil, water and humus are held and nitrogen accumulates. Arboreal ants colonise, lay their eggs and drag seeds that grow in the niche. Insectivorous birds gather to eat ants and eggs; birds also gather and gradually a faunal association grows. Epiphytes are chief components of the vertical structure of our forests.

The trees, shrubs and herbs, the Trinity of environment yield substantial quality of medicinal drugs besides protecting the environment. No amount of administrative and legislative and protective measures to save the country from environmental disaster would be effective and lasting without creation of a dense cover of vegetation all over the country.

Medicinal Herbs and Their Conservation

Lists of medicinal (herbs, shrubs, climbers and trees) have been prepared from the view point of their conservation status. The lists must not be taken as checklists of plants of India for only a few and common species have been selected.The readers are requested to refer to Chaudhuri's work on herbs, shrubs, trees and climbers of India mentioned under referrence which gives a present day view of the plants of India in general and Eastern India in particular.

Conservation related status are given category wise (Numbers indicate status as detailed below):

- Widespread and presently not threatened (Spordic) (S)
- Occur at selected spots and cultivated (C)
- Occur sporadically and require protection (presently safe) (PS).
- Depleted (D).

The study of vegetation reveals that more herbs occur in waste places and in marginal lands and along the edges of various forest types. Evergreen, semi-evergreen wet type of forests have comparatively few dicot species than monocot in terms of numbers of individuals.

Situation of Herb in India

A study undertaken by the author reveals the following:

(i) Few herbs have a wide range of distribution.

(ii) Most of the herbs are confined to forested areas and/or are sporadic indistribution.

(iii) Gregariously occurring herbs are few.

(iv) Moist localities have more number of species than drier and arid localities.

(v) Temperate locations have more of gregarious herbs.

Table 10.1 : Wide distribution of some herbaceous species all over India

Scientific Name	Status	Scientific Name	Status
Sida veronicoefolia (Joka)	S	*Euphorbia hirta (Swadaparui)*	S*
Sida cordifolia (Barlela)	–	*Atylosia scarabioides*	S
Urena sinucata	S	*Mallugc hirta*	S
Acalypha indica (Muktajhuri)	S*	*Elephantopus scaber (Shaydalan)*	S
Evolvulus asinoides	S*	*Polygonum orientale*	S
Solanum nigram	S	*Oxalis acetosella*	S
Solanum indicum	S	*Ageratum conyzoides*	S*
Solanum Torvum	S	*Eclipta alba*	S*
Physalis peruviana	S	*Capparis decidua*	S
Scoparia dulcis	S	*Abrus precatorius (Kunch)*	T
Cenopodium album	S	*Hiptage bengelensis (Madhabilata)*	S
Heliatropium indicum (Srihaslini)	S	*Jasminum grandiflorum (Chameli)*	S
Zomia diphylla	S	*Argyreia speciosa (Rishya gandhri)*	S
Musaenda frondosa (Nagballi)	S	*Marsdenia volubilis (Nach chikni)*	S
Loranthus sp.	S	*Hemidesmus indicus (Anantamul)*	T
Capsella bursapasteris	S	*Cucurbeiaceous climbing shrubs*	S
Cleome isosandra	S	*Oxalis corniculata (Amrul)*	S
Polygala crotalarioides (Nilkanti)	S	*Martynia diandra (Bagh nakhi)*	S
Portulaca oleracea (Barmunia)	S	*Blumea lacera (Kuksim)*	S
Malvastrum sp.	S	*Taraxacum officinale*	S
Vandellia sp.	S	*Sida rohmboidea (Pila Barela)*	S
Rungia parviflora (Piri)	S	*Urena lobata (Ban Okra)*	S
Pogostemon parviflorus	S	*Clitoria turnatea*	S
Amaranthus spinosus (Kanta note)	S*	*Centella asiatica*	S
Amaranthus viridis	S*	*Merremia emerginata*	S
Cannabis sativa	C	*Solanum ferox*	S
Rumex maritimus	S	*Solanum xanthocarpum*	S
Phyllanthus reticulatus (Panjuli)	S	*Solanum surattense*	S
Phyllanthus urinaria (Hazarmani)	S	*Withania somnifera*	T
Cenopodium ambrosioides	S		

contd...

Table 10.1 – *contd...*

Scientific Name	Status	Scientific Name	Status
Picrorrhiza kurtoa	T	*Chrosophora cinera (Tut kukusima)*	S
Abutilon indicum (Paleri)	S	*Lindenbergia urticaefolia (Haldebasanta)*	S
Flemesgia sp.	S	*Oldenlendia corymboisa (Khetkapra)*	S
Trianthema portulacastrum	S	*Wedelia calendula (Bhimraj)*	S
Gangrea maderaspetane (Namuti)	S	*Malachra capitata*	S
Argemone mexicana (Sealkanta)	S*	*Saphora sp.*	S
Cleme gynandra	S	*Oldenlendia herbacea*	S
Polygala chinensis (Muradu)	S	*Celosia argentea*	S
Polycarpea corymbosa	S	*Oxalis corniculata*	S*
Portulaca quadrifida (chota numia)	S	*Rubus sp.*	S
Torenia asiatica	S	*Embelia ribes*	T
Justicia gandurussa (Jagat madan)	PS	*Calamintha sp.*	S
Achyranthus aspera (Apang)	S*	*Achyrathes bidentata*	S
Boerhaevia diffusa (Punarnava)	S*	*Clematis nepalensis (Churahar)*	T
Amaranthus gangeticus (note)	S*	*Clematis smilacifolia*	T
Xanthium strumerium (Ban Okra)	S*	*Thalictrum foliolosum (Mamira)*	T
Rumex dentatus	S	*Nasturtium palustre*	S
Cyanotis sp.	S	*Viola sp.*	S
Phyllanthus simplex	S	*Plantago ovata (Isabgul)*	S
Phyllanthus niruri	S	*Verbascum thapsus*	S
Aerva lanata (Chaya)	S	*Polygonum molle*	S
Chenopodium album (Bestak)	S	*Coptis teeta (Mismitita)*	T
Alocasia indica (Mankachu)	S	*Impatiens sp.*	S
Amorphophallus campanulatus (01)	S	*Potentilla sp.*	S
Commlina bengalensis (Kanachira)	S	*Maesa indica*	S
Acanthus ilicifolius (Harakuchkata)	S	*Brunella vulgaris*	S
Andrographis paniculata (Kalmegh)	T	*Blepharis edulis (Ulangan)*	S
Asterocantha longifolia (Khirok)	S	*Clematis triloba*	S
Biophytum sensetivum (Ban naranga)	S	*Clematis gouriana*	S
Sphaeranthus indica (Mahashrabani)	S	*Corydalis govaniana (Bhutkishi)*	S
		Viola serpens	S
		Geranium sp.	S
		Plantago major	S
		Rubia cordifolia (Monjista)	S
		Polygonum chinense	S

Legend : S = Sporadic; S* = Sporadic, but has concentrated patches; PS = Presently same; – = Not known ; T = Threatened; C= Cultivated

Table 10.1 – *contd...*

Statewise Common and Conspicuous Herb Flora of India (* Sign preceding botanical name by species indicates very useful medicinal plants)	
Andhra Pradesh	
Alysicarpus sp.	*Atylosia sp.*
Crotalaria willdenowiana	Tephrosia purpurea
Vernonia cinera	Alysicarpus monilifer
Ipomoea pes-caprae	Indigofera aspalathoides
Indigofera anneaphylla	* *Boerhavia* sp.
* Crotalaria hirta	* Achyranthes aspera
Xanthium indicum	Cassia occidentalis
Tephrosia tenuis	Crotalaria prostata
Indigofera cordifolia	*Boerhavia diffusa
Acanthospermum hispidium	Celosia argentea
Assam	
Ageratum conyzoides	* Centella asiatica
Inula cappa	Melastoma malabathricum
Mikania micrantha	*Ipomoea* sp.
Alpinia nutans	
Andamans	
Acanthus ilicifolius	*Cucurbita* sp.
Blumea virens	Cassia tora
Desmodium triflorum	Centella asiatica
Jatropha curcas	
Bihar	
* Cassia occidentalis	Vernonia cinerea
* Calotropis procera	* Leucas aspara
* *Cassia tora*	* *Alysicarpus* sp.
* Abutilon indicum	* Tephrosia purpurea
Triadax procumbens	Berleria cristata
* *Desmodium triflorum*	* *Atylosia* sp.
Croton bonplandianum	Argemone maxicana
* Amaranthus spinosus	Borreria articulata
* Abrus precatorius	Corchorus trilocularis
Euphorbia hirta	* Alysicarpus monilifer
Asclepias pseudosansa	Indigofera linifolia
* *Atylosia scaraboeoides*	*Mucuna* sp.
Delhi	
* Achyranthes aspera	Ageratum conyzoides
* Casia occidentalis	* C. tora
* Crotalaria medicaginea	Degera muricata
* Indigogera cordifolia	* Tephrosia purpurea

contd...

Table 10.1 – *contd...*

Rhynchosia capitata	Eclipta prostata
Ipomoea hispida	Borreria articulata
* Tephrosia tenuis	* Justica diffusa
Gujarat	
* Alysicarpus longifolius	Enicostema verticillatum
Heylandia latebrosa	* *Alysicarpus* sp.
Clemome viscosa	Digera muricata
Indigofera cordifolia	Rhynchosia capitata
Vicoa indica	* Eclipta procumbens
* Tridax procumbens	* Borreria articulata
Crotalaria medicaginea	* Barleria cristata
Trichodesma indicus	*Ipomoea* sp.
Andaman	
Ipomea pes-caprae	Cardiospermum halicacabum
Euphorbia thymifolia	Tephrosia purpurea
* Zornia diphylla	Cressa cretica
* Crotalaria burhia	Phynchosia halicacabum
Euphorbia hypericifolia	* Phaseolus aconitifolius
Rhynchosia sp.	*Evolvulus alsionoides*
Commelina benghalensis	*Clitoria ternatea
Vernonia cinerea	*Aerva* sp.
Boerhavia diffusa	
Himachal Pradesh	
* *Atylosia scarabaeoides*	* *Crotalaria* sp.
* Cassia tora	Euphorbia hirta
Jammu & Kashmir	
* Cassia tora	* Tephrosia purpurea
Euphorbia jorta	
Kerala	
* Mimosa pudica	* Desmodium triflorum
Sphaeranthes sp.	* Achyranthes aspera
* Crotalaria alata	* Atylosia scarabaeoides
* Desmodium floribundum	* Moghonia chappar
* *Cassia tora*	* *Crotalaria* sp.
* *Phaseolus* sp.	*Sesbania aculeata*
Karnataka	
* Alysicarpus vaginalis	* Desmodium diffusum
* Indigorera linifolia	* Tephrosia purpurea
Euphorbia hirta	* *Plectranthus* sp.
* Indigofera enneaphylla	* Tephrosia tinctoria
* Clitoria ternatea	* Pueraria phaseoloides

contd...

Table 10.1 – *contd...*

Madhya Pradesh	
Trimufetta bartramia	* Urena lobata
* Alysicarpus vaginalis	* Desmodium triflorum
* Xanthium indicum	* Centella siatica
Caesulia axillaris	*Phaseolus radiatus
* Sesbania bispinosa	* Alysicarpus tetragonolobus
* *Moghania* sp.	* *Cassia tora*
* *Zornia diphylla*	* *Atylosia* sp.
Euphorbia hirta	* Indigofera cordifolia
* Alysicarpus rugosus	* Borreria articulata
* Corchorus trilocularis	* Indigofera linifolia
Maharashtra	
* Crotalaria retusa	* Aerva lanata
* Alysicarpus vaginalis	* Cassia tora
* Indigofera cordifolia	* I. linifolia
Phaseolus trilobus	Polygala erioptera
* Tephrosia purpurea	Tephrosia tenuis
* *Amaranthus* sp.	* *Desmodium diffusum*
Xanthium indicum	* Crotalaria prostata
* Desmodium triflorum	Heylandia latebrosa
* Crotalaria prostata	* Desmodium triflorum
Heylandia latebrosa	* rotalaria sericea
* Indigofera glandilosa	* Achyranthea aspera
* *Atylosia* sp.	* *Crotalaria alata*
* C. burhia	* Alysicarpus rugosua
* Alysicarpus longifolius	* Sesbania bispinosa
* Crotalaria juncea	Atylosia scarobaeoides
Argemone mexicana	* Crotlaria vestita
* Indigofera hebepetata	* Alysicrpus tetragonolobus
Manipur	
* *Crotalaris* sp.	* *C. medicaginea*
* *Desmodium* sp.	* *Indigofera enneaphylla*
Orissa	
* Cassia occidentalis	Croton bonplandianum
* Indigofera enneaphylla	* Desmodium triflorum
* Cassia tora	* Tridax procumbens
* *Crotalaria* sp.	*Triumfetta bartramia*
Ageratum conyzoides	Eranthemum Purpurascens
* Acyranthes aspera	* Alysicarpus vaginalis

contd...

Table 10.1 – *contd...*

Punjab	
* *Achyranthes aspera*	* *Alysicarpus* sp.
Rhynchosia capitata	* Indigofera linifolia
* Vicoa indica	* Casia tora
* Boerhavia repens	* Crotalaria medicaginea
Xanthium indicum	* Tephrosia purpurea
* Atylosia scarabaeoides	Vernonia cinerea
Chenopodium album	Euphorbia hirta
Trigonella polycerata	* *Artemisia* sp.
* Indigofera cordifolia	* I. linifolia
Rajasthan	
* Achyranthes aspera	* Alysicarpus longifolia
* Cassia tora	* Crotalaria medicaginea
* Indigofera enneaphylla	* I. linifolia
* I. cordifolia	* Tephrosia purpurea
* T. tenuis	* Xanthium indicum
* Amaranthus blitum	* Crotalaria burhia
* C. albida	* Eclipta alba
Alysicarpus vaginalis	Argemone mexicana
* Boerhavia repens	* Crotalaria filipes
* Tridax procumbens	Justicia diffusa
* Aerva tomentosa	* Crotalaria umbellata
Heylandia latebrosa	* Phaseolus radiatus
* Boerhavia diffusa	Dicoma tomentosa
* Mimosa pudica	* Zorina diphylla
Tribulus terrestris	* Desmodium rotundifolium
* Alternanthera echinata	
Tamil Nadu	
Rhynochosia Psuedo-cajan	* *Tephrosia* sp.
Xanthium indicum	* Indigofera enneaphylla
* *I.linifolia*	*Stylosanthes* sp.
* Atylosia scarbeoides	* Desmodium floribundum
* Tephrosia purpurea	Stylosanthus fruticosa
* Alysicarpus vaginalis	Croton banpladianum
Euphornbia thymifolia	* Leucas hyssopifolia
* Indigofera trifoliata	* Achyranthes aspera
* *Atylosia* sp.	* *Crotalaria prostrata*
* Indigofera viscosa	* Pogstemon mollis
* Mimosa pudica	Polygala crotalarioides
Euphorbia trigona	* Alysicarpus monilifer
* *Indigofera trigonelloides*	*Mollugo* sp.

contd...

Table 10.1 – *contd...*

Stachytrapheta indica	* Jatropha curcas
Ageratum conzyzoides	* Sida acutangula
Rhynchosia rufescens	
Tripura	
Carex condensata	* Hyptis brevipes
* Desmodium trifolium	* Ocimum sanctum
* Atylosia scarabaeoides	* Tridax procumbens
Uttar Pradesh	
* *Alysicarpus* sp.	* *Cannabis sativa*
* Euphorbia hirta	* Leuças aspera
* *Phyllanthus maderaspatensis*	* *Atylosia* sp.
* *Vicoa indica*	* *Atylosia* sp.
Heylandia latebrosa	* Indigofera linifolia
* Tephrosia villosa	* Cassia tora
* Crotalaria medicaginea	* Desmodium triflorum
Vernonia cinerea	* Tridax procumbens
* Zornia diphylla	* Xanthium indicum
* Aerva lanata	* Alysicarpus monilifer
Boerhavia repens	Justicia diffusa
Cassia occidentalis	* *Chenopodium* sp.
Spiraea canescens	*Stellaria* sp.
Stellaria decumbens	* Indigofera heterantha
* Polygonum cirrhifolium	
West Bengal	
* *Centella asiatica*	* *Moghania* sp.
* Leucas aspera	Vernonia cinerea
Argemone mexicana	* Atylosia scarabaeoides
* Cassia tora	* Clitoria ternatea
Croton bonplandianum	* Datura metel
Digera muricata	Euphorbia hirta
* Leucas aspera	Mikania micrantha
* Desmodium triflorum	* Alysicarpus monilifer
Borreria acticulata	* Vicoa indica
* Tephrosia purpurea	* Tridax procumbens
* Acanthus ilicifolius	* Sesuvium Portulacastrum
* Cassia occidentalis	Phyla nodiflora
* Tephrosia villosa	Ipomoea pes-caprae
* Achryanthes aspera	Triflorum repens
* *Polygonum* sp.	*Inula cappa*
* *Canabis sativa*	* *Pueraria* sp.

Chapter Eleven
Uses of Some Medicinal Plants of Selected Areas and Their Status

The aim and objective of this treatise is not enumerate the qualities of medicinal plants of the country. At the same time it is an imperative necessity to aware the people about the efficacy of some of the innumerable drug plants of the country.

As such the author has pickup some common plants of multifacet qualities and of some selected areas of the country and enumerated various uses.

The areas are:

- Medicinal plants of Arid and Desert areas.
- Plants of Sundarbans mangrove.
- Bankura district of West Bengal.
- Plants of Satpura plateau.

The status of each species of the areas has been mentioned.

Medicinal Properties of Some Selected Plant Species

This subject has been dealt with very elaborately by all the frontline authorities on the subject. As such the information presented in their work is not being repeated. Sibakali Bhattacharyya has dealt with various aspects of each sp. In his work (in 10 volumes) entitled "Chiranjib Vanusodhi", the author seems to cover this subject elaborately. Works of other authorities also give similar information.

In order to have a broad idea about the subject the present author has enumerated some information in diagrams for easy access to the subject by ordinary readers. These are some diagrammatic and some non-diagramatic presentations of several very important and common plants.

There is perhaps a lacuna in the work of the frontline workers which is the omission in their work about mentioning the "remedy of a disease by a specific plant" (mention has been made of several plants as medicinal agents for particular disease). So a patient does not know what plant would be best to remedy the disease he are suffering from. The present authors are not sure if a plant or plants can be pinpointed to remedy a particular diseases.

Table 11.1

Scientific name	Common name	Medicinal Uses
Bombax ceiba	Semul	Dried stem and bark used in impotency, Gastro-intestinal disorder, Cough and female diseases.
Butea monosperma	Palash	Seeds and gum used in eradication of Intestinal worm, Sexual impotency, Restoration of menstrual cycle, Prevent pregnancy.
Mimosa pudica	Lajwati	Seed used in sexual impotency checking flow of seman through urine, Strong anthelmentic.
Allum satiuum	Rasun	Dyspepsia, hiecough, Infantile corvulsions, Nervous affections.
Basella rubra	Pui Sag	Genitournary tract, Headache, Insomnia.
Rauwolfia serpentina	–	In bites of poisonous reptiles and insects, corneal opacity, Cholders in painful bowel affection, High blood pressure, Insomnia.
Piper nigram	Golmorich	Anthelmentic, Asthma, Throat disease, Piles, Night blindness, Carminative, Aphrodisiac, Purgative, Toothache.
Euphorbia hirta		Dysentery, Colic pain, Bowel complaint, Increase flow of mith in women, Diarrhoea.
Emblica officinalis	Amloki	Carminative, Laxative, Tonic, Antipyretic, Biliousness, Urinary Discharge, Leprosy, Anaema, Opthalmia, Aphrodasiac Asthma. Bronchitis.
Jatropha curcas	Bharenda	Anthelmentic, Chronic dysentery Urinary discharge, Anaema, Fistula,

contd....

Table 11.1 contd....

Scientific name	Common name	Medicinal Uses
		Bad gum, Eczema, Wing worm.
Cynodon dactylon	Durba graze	Dysentery, Wound, Pyrrhoea, Piles.
Phoenix sylvostris	Khejur	Bronchitis, Haemoptysis, Worm, Cough.
Piper cubeca	Kababchini	Dysentery, Asthma, Gout.
Aegel marmelos	Bel	Cough, Cold, Dysentery, Scurvy, Typhoid, constipation.
Santalum album	Chandan	Fever, Cough, Gonorrhcea, Epilepsy, Pox, Whooping Cough. Stomachic, tonic anthelminitic, Kalazar.
Andrographis paniculata	Kalmegh	Bronchitis, Leprosy, Blood Purifier, Heart trouble, Asthma, Vomitting, Loss of memory, Leucoderma, Diuretic Bronchitis, Asthma, Gonorrhoea.
Vites negunda	Nisindha	Astingent, stomachic, Anthelmintic, for growth of hair. Eve disease. Leucoderma. Bronchitis, Asthma.
Ocimum basilicum	Bhutulsi	Stomachic, Anthelmentic, Cough, Gonorrhoea, Diarrhoea, Scorpoon sting, Snake bite.
Ocimum sanctum	Tulsi	Stomachic, Antipyretic, Heart disease, Leucoderma, Bronchitis expectorant.
Curculigo orchiodes	Shyam mushli	Sexual impotency, Veneral and Urinary diseases Asthma, Jaundice, (Dried tuberous roots).

Some Common Diseases and Their Remedy from Plants

Many plants and their parts are source for remedy of a single disease and also many more diseases. It could not be ascertained from the works of the researchers as to which species or what parts are most effective in the remedy.

Some More Diseases and Plants Used as Remedy

Anti-coagulant:

Justica adhatoda, Allium sativum, Azadiachta indica, Camellia sinensis.

Cardiovascular (Hypertensive and Cardio tonics):

Aegle marmelos (root, bark), Allium pepa, Allium sativum, Arachis hypogea,Asclepias curassavica, Cururbita maxima, Piper aurantiacum, Rauwolfia serpentina (root), Terminalia arjuna (bark).

Aphrodisiac :

Allium sepa, Allium sativum, Aasparagus racemosus, Glycyrrhiza sp., Hibiscus-rosa-sinensis Myristica fragans, Mucuna puriens, Strychonos nux-vomica, Tinospora cordifolia.

Antiarthritis, Antiinflammatory :

Acorus calamus, Allium sativum, Boswellia sarrata, Calophyllum innophyllum, Capsicum sp. Curcuma longa, Eclipta abla, Solanum nigrum, Withania somnifera, Xanthium strumerium, Zingiber officinale.

Antiasthmatic :

Acalyha indica, Acorus calamus, Justicia adhatoda, Albizzia lebbek, Ephedra sp., Ocimum sanctum, Vitex negundo, Terminalia belerica, Piper longum, Saussurea lappa, Picrorhiza Kurroa, Lobelia sp., Hedycium sp.

Antidiabetic :

Aegle marmelos, Allium sativum, Anona squamosa, Asteracantha longifolia, Azadirachta indica, Blumea sativum, Cinnamomum tamala, Crurcuma longa, Emblica officinalis, Gymnema sylvestre., Momordica charantia, Zingiber officinale, Swertia chirata, Syzygium cumini (seed), Musa paradisiaca.

Bronchitis (Respiratory disorder) :

Picrohiza kurroa (Kutki)-dried rihizome; Zingiber officinale (Adark)-dried rhizome, Myrica nagi (Kaifal)-Dried root & bark; piper longum (bari papal)-Fruits; Glycyrrhiza glabra-(Muleltee)-dried roots; Pistacia integerrima (Kakrusungi)-Galls; Desmondium gangeticum (Salparni)-Leaf, stem; Mesua ferrea (naG-Kashar)-Fruit, Seed; Withania somnifera (Anwagandha)-Dried roots.

Liver disease :

Acacia catechu, Azadirachta indica, Andropogon paniculata, Asteracantha longifolia, Andrographis paniculata, Allium sativum. Boerhaavia diffusa, Cynodon dactylon, Capparis spinosa (Climbing shrub); Emblica officinalis.

Memory stimulant :

Acorus calmus, Bacopa monnrieri, Centrella asiatica, Celastrus paniculata (Climber), Glycyrrhiza glabra.

Hyper acidity :

Acorus calamus, Asparagus racemosa, Emblica officinalis, Glycrrhiza glabra, Zingiber officinale.

Viral Diseases:

Among the virses that infect human being, most common are HIV (Human Immunodeficiency Virus), HBV (Hepatitis-B Virus), IV (Influenza Virus), PV (Polio Virus), MV (Measles Virus) and some other viruses. Of these HIV cause AIDS (Acquired Immuno Deficiency Syndrome), appears lethal. An HIV attacts four lymphocytes of the human immune system. Plants Kill HTV directly are papaver (Opium), Wild snake, Periwinkle, Croton, Anemone, Gourd, Licorice (Jastimadhu, Castor (Aranila), Pine cone ,King Bitter (Kalmesh)etc. Plants having components to interfere replications.

Galactagogue :

Plants containing several compounds which indue milk production in greater volume and secretio of the same in the human beings are termed galactagogues.

Number of plants species recognized as experimentally proven species are *Alstonia Schloaris* (Chattni),Oat, Fig. (Dummor), Fenugreek (Methi), Lettace (Jeera), Barley, Onion.

Various Qualities of Medicinal Plants

No elaborate information has been presented. The issue is very elaborate but the subject under discussion has a limited field. However some information are presented in the following paragraphs some of which are plants used in poliomyelites, Massage oil, specific diseases of women, beauty and care and also on ayurvedic uses.

It is just a broad view of the subject.

- **Creative value of some widely known plants.**

Adhatoda vasica	- Expectorant and antiasthmatic.
Aegle marmelos *Plantago ovata* *Ailanthes malabarica*	- Chronic diarrhoea & dysentery.
Bacopa monnieri	- For memory
Centella asiatica	- For intelligence

Melia azadirachta	- Antiperiodic and many other diseases.

(*Source* : Ind. For. April 2004 : A.K. Bhattacharjee)

- **Suggested for poliomyclites-Oral drug.**

Asparagas racemosus	- Dried roots
A. dscendens	- "
Bombax ceiba	- Dried bark
Mucuna pariens	- Seeds
Pueraria tuberosa	- Dried tuber
Withania somnifera	- Dried roots
Butea monospermat	- Gums
Orchis latifolia	- Dried rhizome
Curculigo orchioides	- Dried roots
Salaginella bryopteris	- Leaves
Cassia fistula	- Fruits and seeds
Parmelia tinctorium	- (Lichen)-Thallus
Tinospora cordifolia	- Dried stem
Tribulus terrestris	- Dried fruits
Bambusa bamboos	- Secretions
Cotula anthelmoides	- Stem
Gardenia gummifera	- Gum

- **Herbal massage oil-Curative to Children.**

Tribulus terrestris	- Dried fruits
Pedalium sp.	- Dried fruits
Cotula anthemoides	- Dried stem and leaves
Gardenia gummifera	- Gum
Colchicum tuteum	- Dried corn
Myrica nagi	- Dried root bark
Embelia ribes	- Dried fruits
Mesua ferrea	- Flower, bud and fruits
Onsoma echoides	- Dried flower
Myristica fragrans	- Dried fruit
Carum coptieum	- Seeds
Odina woodier	- Gums.

- **Species recommended for Women's Menorrhagia**

Symplocos racemosa	-	Dried stem bark
Cocculus villosus	-	Leaves
Bombax ceiba	-	Stem bark
Rhus coriaria	-	Galls
Mesua ferrea	-	Fruits and seeds
Myrica nagi	-	Root bark
Picrorhiza kurroa	-	Dried rhizome

- **Species recommended for women's Leucorrhoea**

Symplocos racemosa	-	Dried stem bark
Paeraria tuberosa	-	Tuber
Butea monosperme	-	Gum
Sida acuta	-	Seeds
Withania somnifera	-	Roots

- **Plants used in Beauty and Care**

Argemone maxicana	Aloe vera
Musa sp.	Bixa orellana
Medicago sp.	Terminalia arjuna
Prunus sp.	Asparagus racemosus
Rubes sp.	Terminalia bellirica
Ribes sp.	Allium Sepa.
Rumex sp.	Vitis sp.
Viola sp.	Melilotus sp.
Plantago sp.	
Glycine sp.	
Zyziphus sp.	

- **Some Ayurvedic medicinal plants.**

Acacia cubeba	*Aristolochia Sp.*	*Bauhinia Sp.*
A. farniciana	*Artocarpus Sp.*	*Berberis Sp.*
Acoras calamus	*Aspoaragus racemosa*	*Saxifraga Sp.*
Aloe vera	*Azadirachata indica*	*Bombax ceiba*
Alpinia galanga	*Bacopa Sp.*	*Calotropis Sp.*

Altingia excelsa	*Balanites Sp.*	*Calophyllum mophyllum*
Areca catechu	*Balliospermum Sp.*	*Bachanania lanzan*
Argimone maxicana	*Atropa acuminata*	*Butea monosperma*
Carum Sp.	*Coptis teeta*	*Tribullus terrestris*
Carica occidenlalis	*Cordia myxa*	*Taxus baccata*
Cassia occidenlalis	*Costus specious*	*Terminalia arjuna*
C. tora	*Corton oblongifolius*	*Rauwolfia serpentina*
C. fistula	*C. tiglium*	*Sida cordifolia*
C. absus	*Curculigo orchioides*	
Chenopodium Sp.	*Curcuma longa*	
Cinnamomum zeylanicum	*C. zeodaria*	
	Cynodon dactylon	
Cissus vuadrangularis	*Cyperus rotundus*	
Calotropis Sp.	*Dillenia indica*	
Clitorea ternate	*Digitalis purpurea*	
Coccinia indica	*Dioscorea bulbifera*	
Eclypta alba	*Vernonia cinera*	
Zizyphus jijuba		
Zanthophyllum alatum		

- **Species having High valued domestic need.**

Emblica officinalis (Amla), *Withiana so nifera*, *Allicum sativum* (Lahsun), *Curcuma longa* (Haridra), *Picrorhiza carroa* (Kutki), *Tinospora cordifolia* (Gulduchi), *Ocimum sp.* (Tulsi), *Zingiber officinalis* (Adark), *Azadirachta indica* (Neem), *Convolvulus pluricaulis/ Evolvulus alsinoides* (Sankhapushpi), *Asparagas racemosus* (Shatavari), *Commiphora mukul* (Guggul), *Aconitum ferox* (Valsanbh), *Garcinia cambagia* (Kokum), *Plantago ovata* (Isabgul), *Holarrhena pubescens* (Kutaya), *Androagraphis paniculata* (Kalmegh), *Saussura lappa* (Kushth), *Saraca asoca* (Asoka), *Justicia adhatoda* (Vasak).

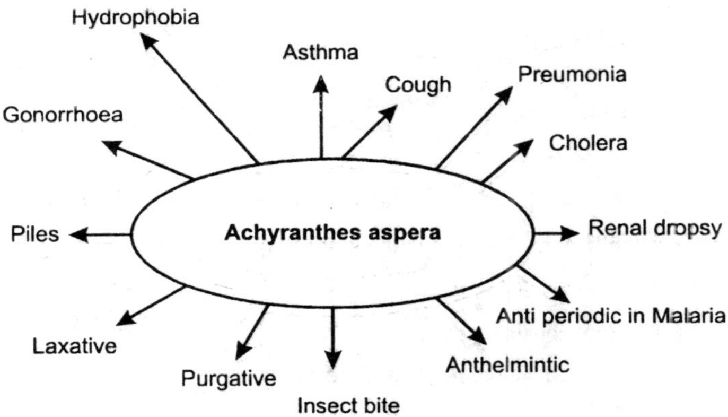
Hydrophobia
Asthma
Cough
Preumonia
Gonorrhoea
Cholera
Piles
Achyranthes aspera
Renal dropsy
Anti periodic in Malaria
Laxative
Anthelmintic
Purgative
Insect bite

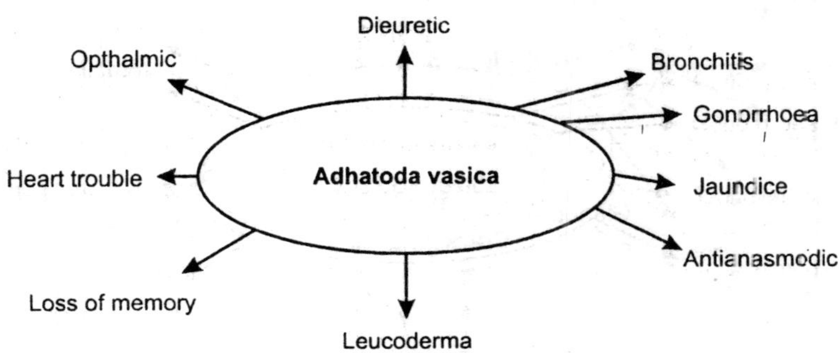
Dieuretic
Opthalmic
Bronchitis
Gonorrhoea
Heart trouble
Adhatoda vasica
Jaundice
Antianasmodic
Loss of memory
Leucoderma

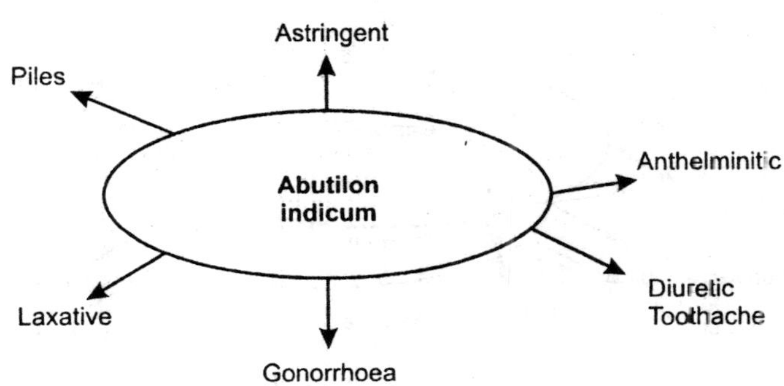
Astringent
Piles
Anthelminitic
Abutilon indicum
Diuretic
Toothache
Laxative
Gonorrhoea

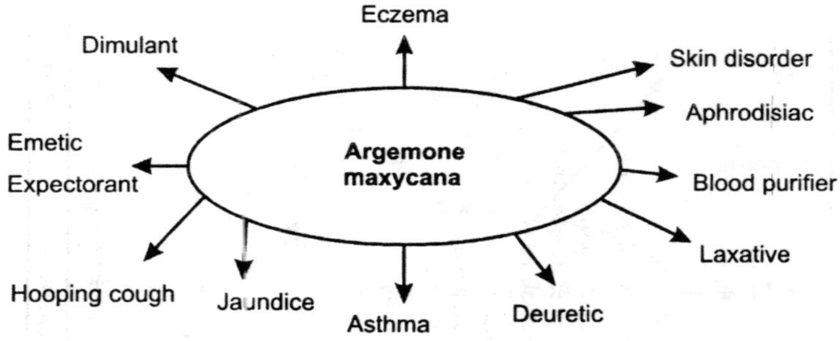
Eczema
Dimulant
Skin disorder
Aphrodisiac
Emetic
Expectorant
Argemone maxycana
Blood purifier
Laxative
Hooping cough
Jaundice
Asthma
Deuretic

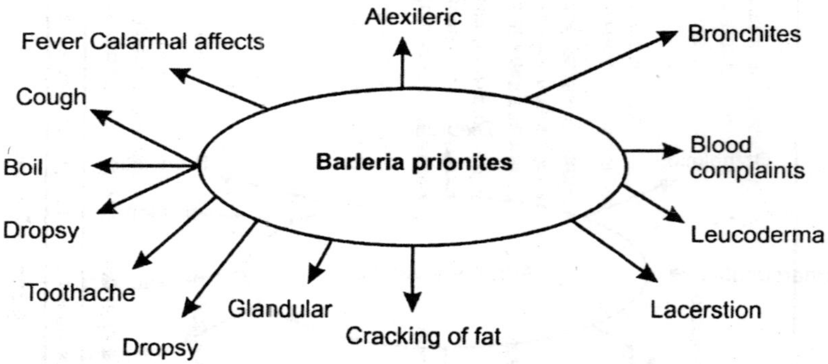
Alexileric
Fever Calarrhal affects
Bronchites
Cough
Boil
Barleria prionites
Blood complaints
Dropsy
Leucoderma
Toothache
Glandular
Lacerstion
Dropsy
Cracking of fat

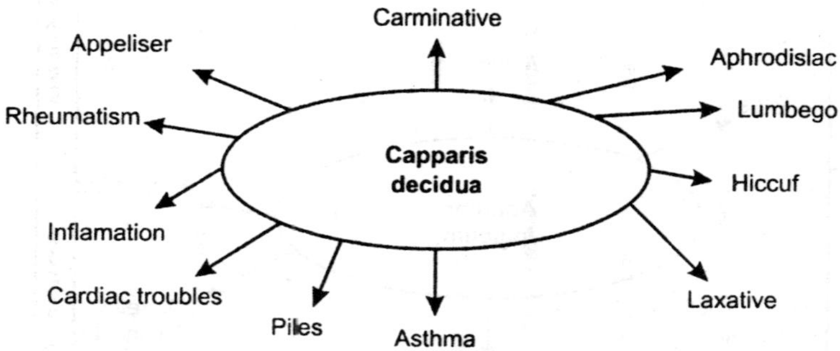
Carminative
Appeliser
Aphrodislac
Rheumatism
Lumbego
Capparis decidua
Hiccuf
Inflamation
Cardiac troubles
Laxative
Piles
Asthma

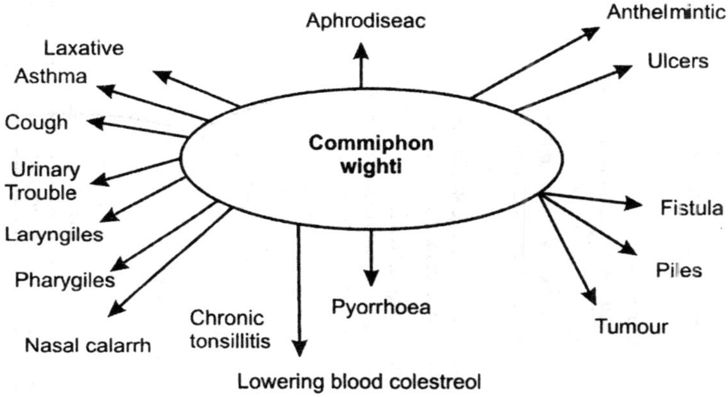
Commiphon wighti
Aphrodiseac
Anthelmintic
Laxative
Asthma
Ulcers
Cough
Urinary Trouble
Fistula
Laryngiles
Piles
Pharygiles
Pyorrhoea
Tumour
Chronic tonsillitis
Nasal calarrh
Lowering blood colestreol

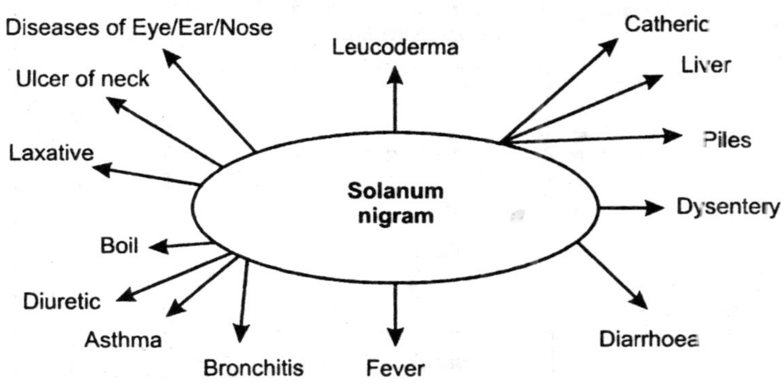
Solanum nigram
Diseases of Eye/Ear/Nose
Leucoderma
Catheric
Ulcer of neck
Liver
Laxative
Piles
Dysentery
Boil
Diuretic
Asthma
Bronchitis
Fever
Diarrhoea

Loss of Qualities of Some Medicinal Plants of India

(Diagrammatic Presentation)

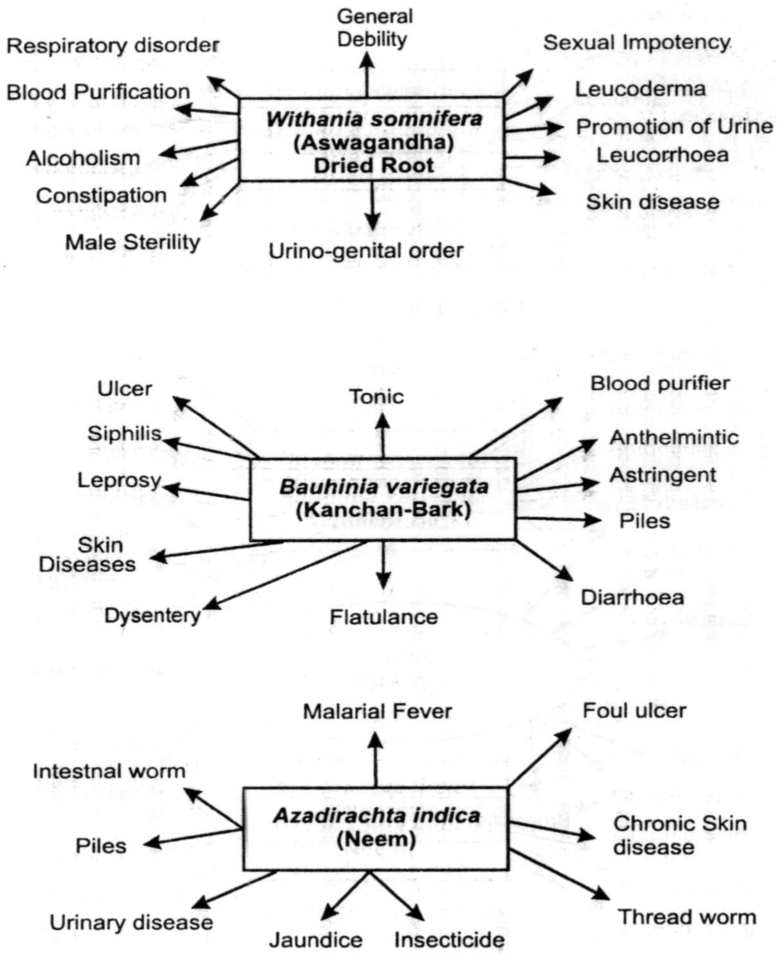

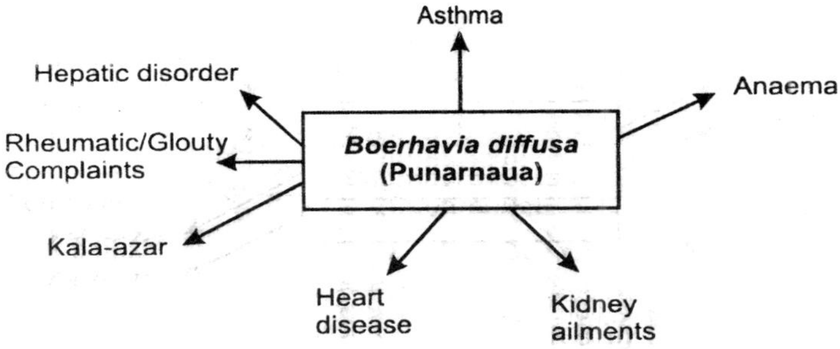
Asthma
Hepatic disorder
Anaema
Rheumatic/Glouty Complaints
Boerhavia diffusa (Punarnaua)
Kala-azar
Heart disease
Kidney ailments

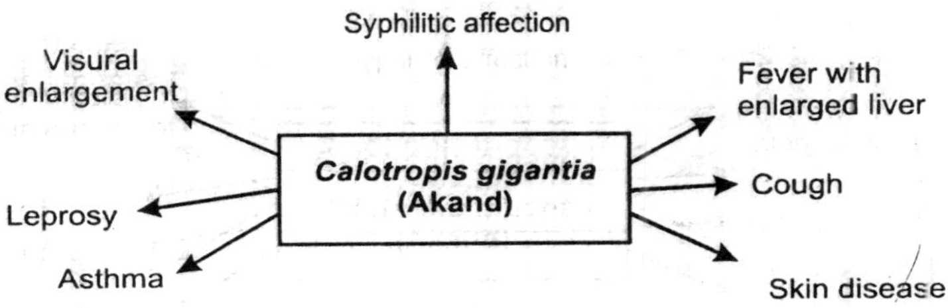
Syphilitic affection
Visural enlargement
Fever with enlarged liver
Calotropis gigantia (Akand)
Cough
Leprosy
Asthma
Skin disease

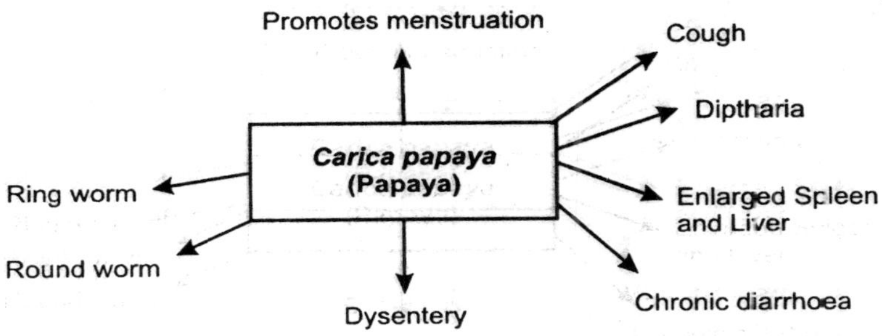
Promotes menstruation
Cough
Diptharia
Carica papaya (Papaya)
Ring worm
Enlarged Spleen and Liver
Round worm
Chronic diarrhoea
Dysentery

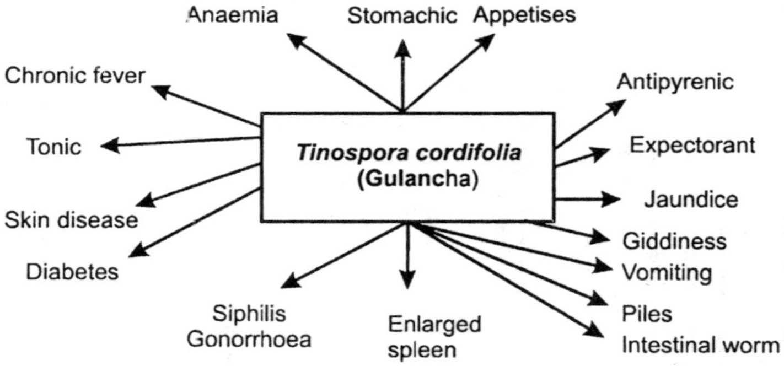
Anaemia
Stomachic
Appetises
Chronic fever
Antipyrenic
Tonic
Tinospora cordifolia (Gulancha)
Expectorant
Jaundice
Skin disease
Diabetes
Giddiness
Vomiting
Siphilis Gonorrhoea
Enlarged spleen
Piles
Intestinal worm

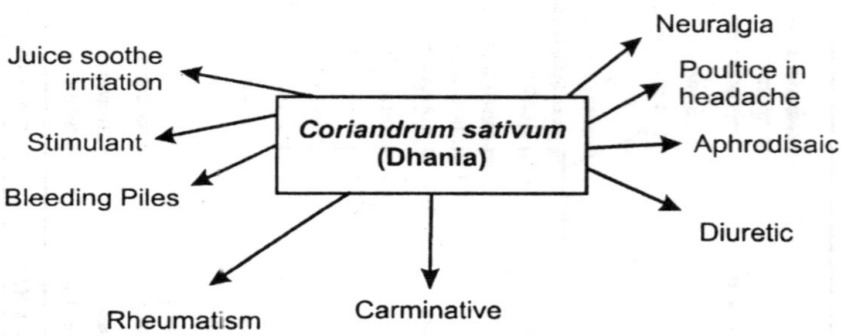
Neuralgia
Juice soothe irritation
Poultice in headache
Stimulant
Coriandrum sativum (Dhania)
Aphrodisaic
Bleeding Piles
Diuretic
Rheumatism
Carminative

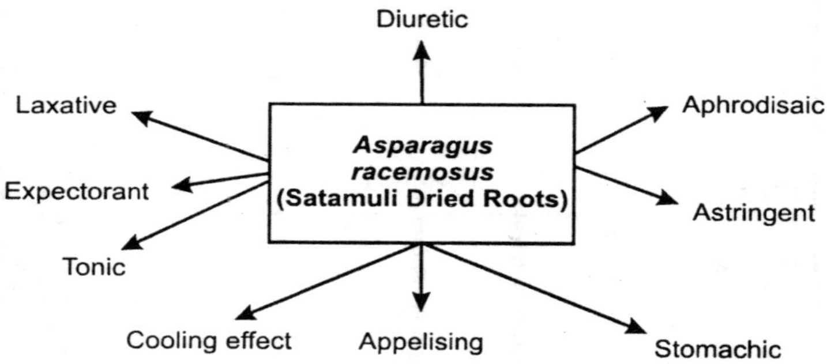
Diuretic
Laxative
Aphrodisaic
Asparagus racemosus (Satamuli Dried Roots)
Expectorant
Astringent
Tonic
Cooling effect
Appelising
Stomachic

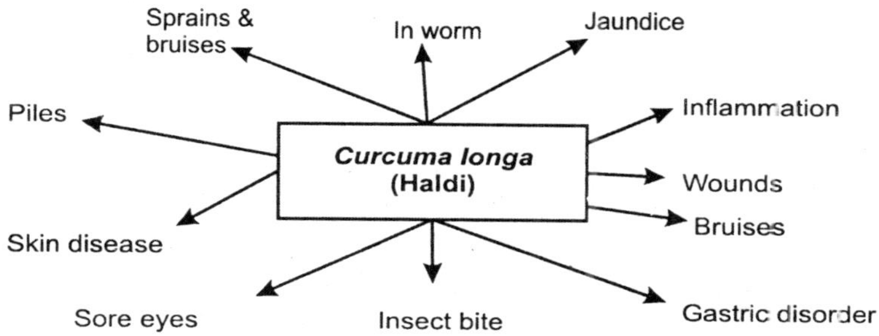
Sprains &
bruises
In worm
Jaundice
Piles
Inflammation
Curcuma longa
(Haldi)
Wounds
Bruises
Skin disease
Sore eyes
Insect bite
Gastric disorder

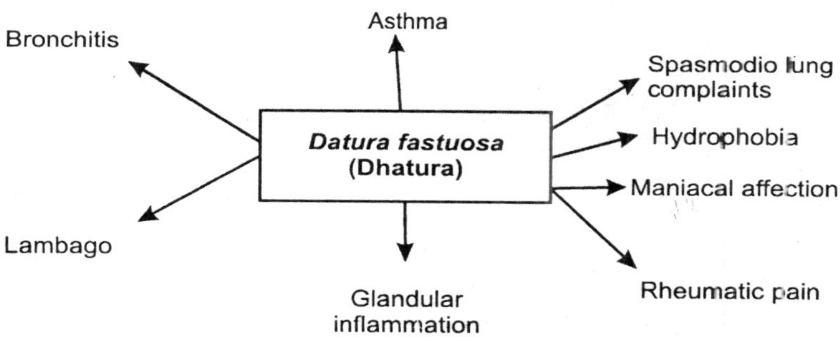
Asthma
Bronchitis
Spasmodio lung
complaints
Datura fastuosa
(Dhatura)
Hydrophobia
Maniacal affection
Lambago
Rheumatic pain
Glandular
inflammation

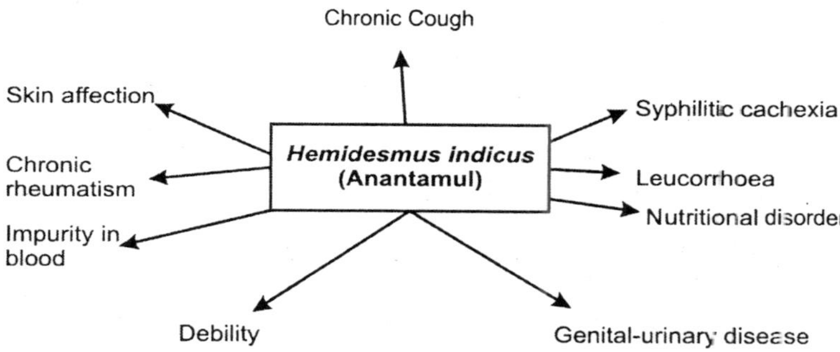
Chronic Cough
Skin affection
Syphilitic cachexia
Hemidesmus indicus
(Anantamul)
Chronic
rheumatism
Leucorrhoea
Nutritional disorder
Impurity in
blood
Debility
Genital-urinary disease

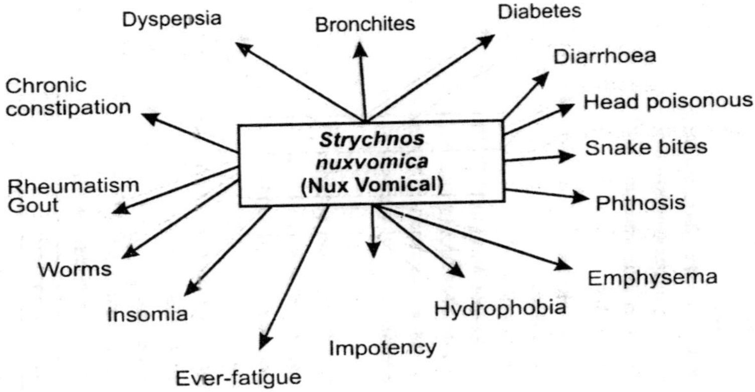
Dyspepsia
Bronchites
Diabetes
Diarrhoea
Chronic constipation
Head poisonous
Strychnos nuxvomica (Nux Vomical)
Snake bites
Rheumatism Gout
Phthosis
Worms
Emphysema
Insomia
Hydrophobia
Impotency
Ever-fatigue

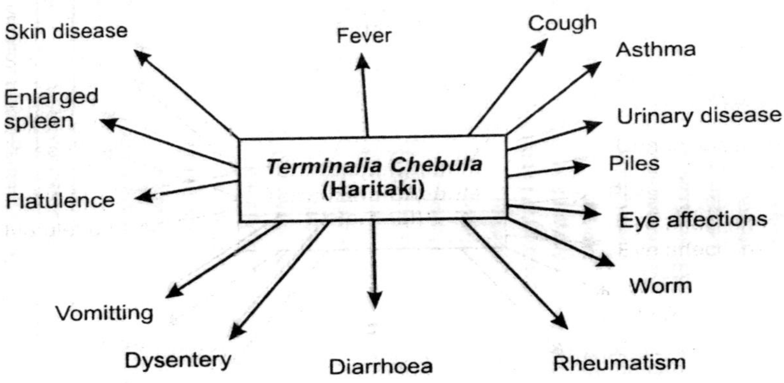
Skin disease
Fever
Cough
Asthma
Enlarged spleen
Urinary disease
Terminalia Chebula (Haritaki)
Piles
Flatulence
Eye affections
Worm
Vomitting
Dysentery
Diarrhoea
Rheumatism

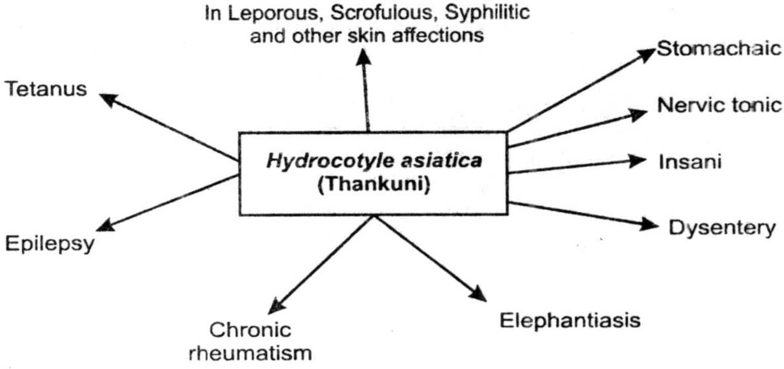
In Leporous, Scrofulous, Syphilitic and other skin affections
Tetanus
Stomachaic
Nervic tonic
Hydrocotyle asiatica
(Thankuni)
Insani
Dysentery
Epilepsy
Chronic rheumatism
Elephantiasis

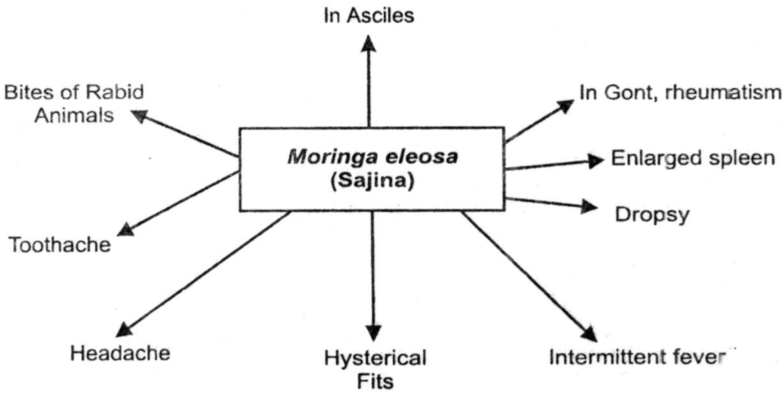
In Asciles
Bites of Rabid Animals
In Gont, rheumatism
Moringa eleosa
(Sajina)
Enlarged spleen
Dropsy
Toothache
Headache
Hysterical Fits
Intermittent fever

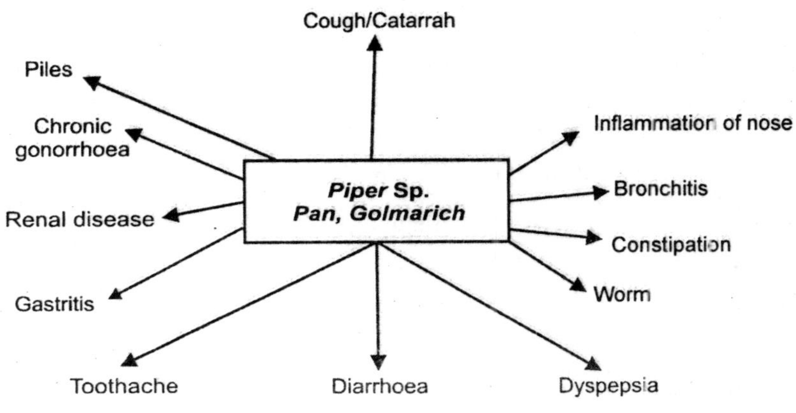
Cough/Catarrah
Piles
Chronic gonorrhoea
Inflammation of nose
Piper Sp.
Pan, Golmarich
Bronchitis
Renal disease
Constipation
Worm
Gastritis
Toothache
Diarrhoea
Dyspepsia

Some Common Diseases and Their Remedy from Plants

Many plants and their parts are source for remedy of a single disease and also many more diseases. It could not be ascertained from the works of the researchers as to which species or what parts are most effective in the remedy.

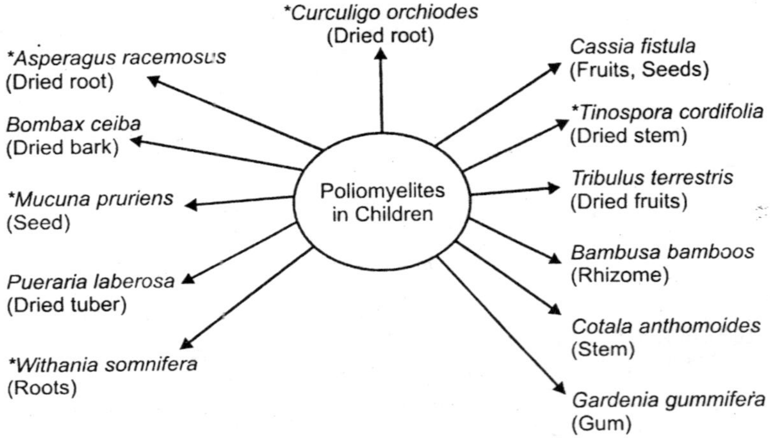

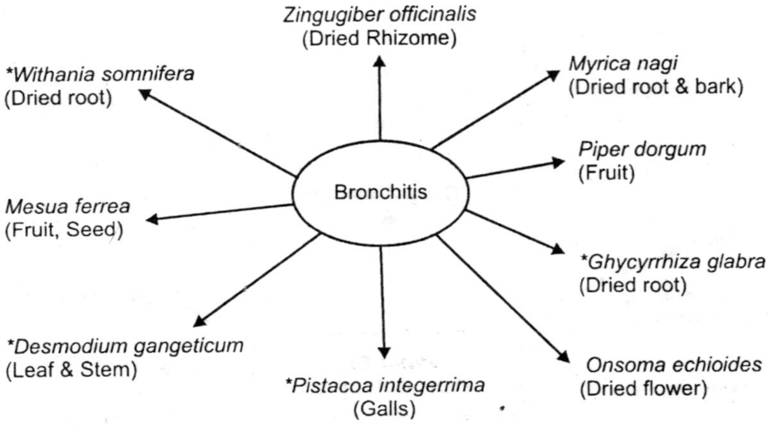

* Not common, rare

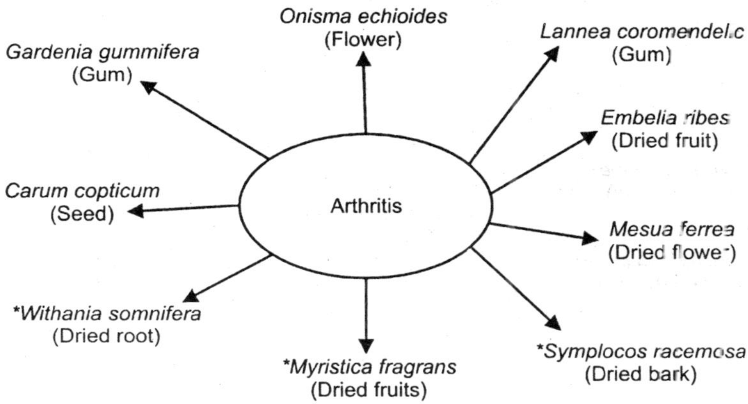

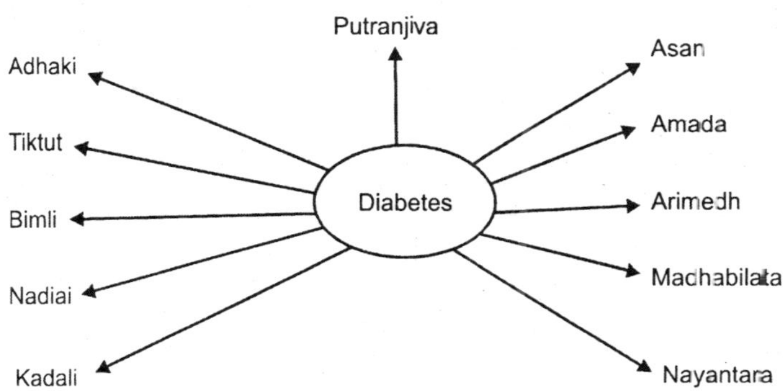

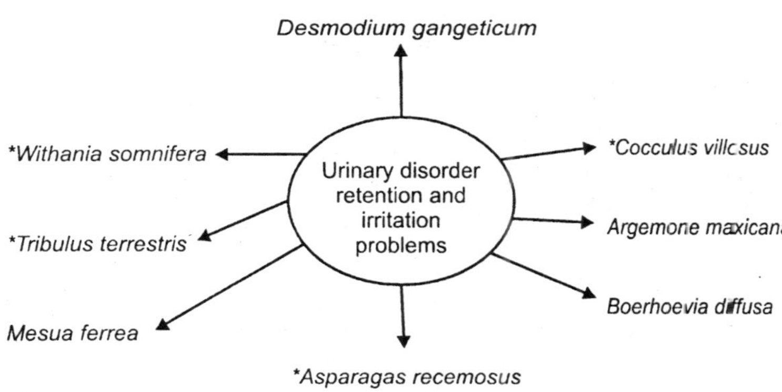

* Not common, rare

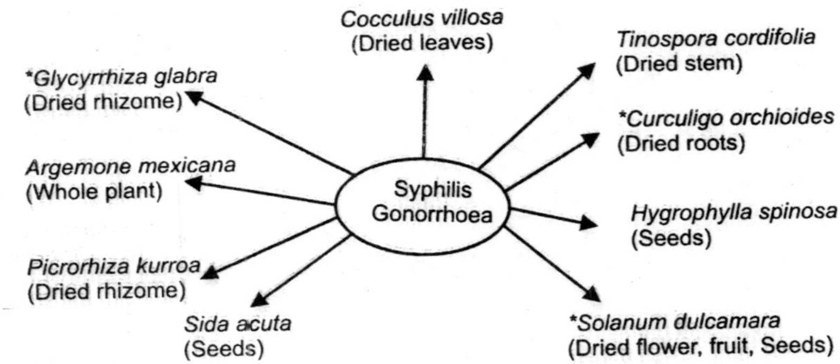
Syphilis
Gonorrhoea
Cocculus villosa
(Dried leaves)
Tinospora cordifolia
(Dried stem)
*Glycyrrhiza glabra
(Dried rhizome)
*Curculigo orchioides
(Dried roots)
Argemone mexicana
(Whole plant)
Hygrophylla spinosa
(Seeds)
Picrorhiza kurroa
(Dried rhizome)
Sida acuta
(Seeds)
*Solanum dulcamara
(Dried flower, fruit, Seeds)

Birth Control
Lawsonia alba (Mehandi) – Fresh Leaves – Common in villages
Tectona grandis (Segun) – Fruits, tender leaves – Common all over India
Butea monosperma Palash) – Fresh seeds – Common in all dry forests
Carica papaya (Papaya) – Fresh seeds – Common all over India

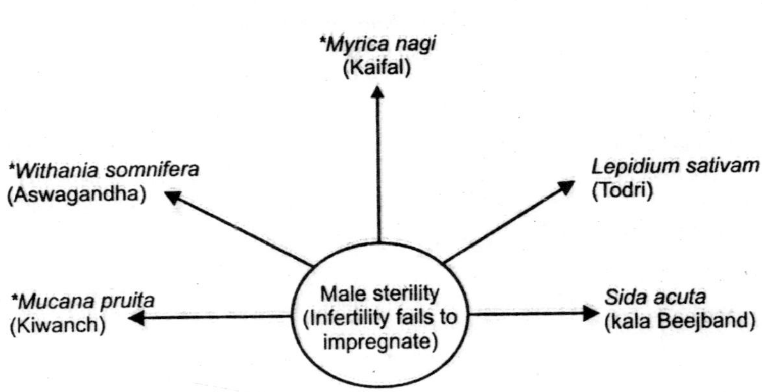
Male sterility
(Infertility fails to
impregnate)
*Myrica nagi
(Kaifal)
*Withania somnifera
(Aswagandha)
Lepidium sativam
(Todri)
*Mucana pruita
(Kiwanch)
Sida acuta
(kala Beejband)

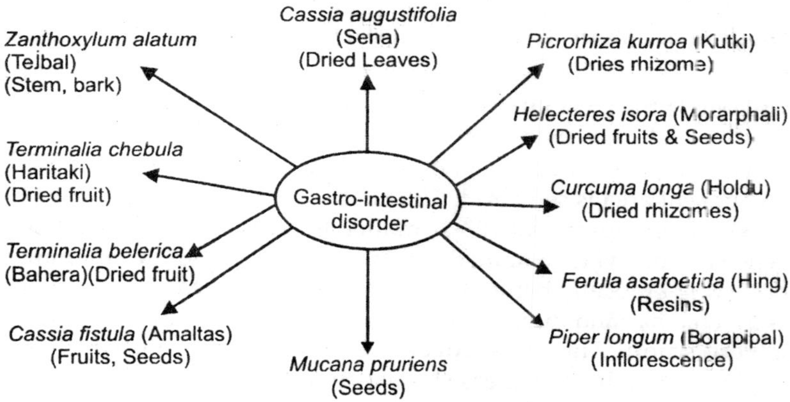
Gastro-intestinal disorder
Cassia augustifolia (Sena) (Dried Leaves)
Zanthoxylum alatum (Tejbal) (Stem, bark)
Picrorhiza kurroa (Kutki) (Dries rhizome)
Helecteres isora (Morarphali) (Dried fruits & Seeds)
Terminalia chebula (Haritaki) (Dried fruit)
Curcuma longa (Holdu) (Dried rhizomes)
Terminalia belerica (Bahera)(Dried fruit)
Ferula asafoetida (Hing) (Resins)
Cassia fistula (Amaltas) (Fruits, Seeds)
Piper longum (Borapipal) (Inflorescence)
Mucana pruriens (Seeds)

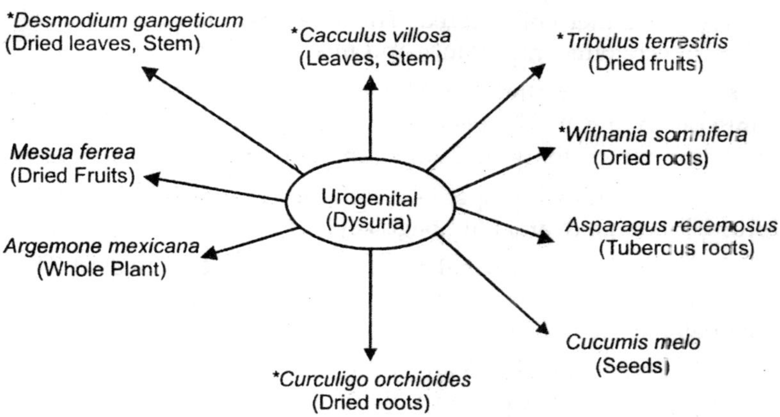
Urogenital (Dysuria)
*Desmodium gangeticum (Dried leaves, Stem)
*Cacculus villosa (Leaves, Stem)
*Tribulus terrestris (Dried fruits)
*Withania somnifera (Dried roots)
Mesua ferrea (Dried Fruits)
Asparagus recemosus (Tubercus roots)
Argemone mexicana (Whole Plant)
Cucumis melo (Seeds)
*Curculigo orchioides (Dried roots)

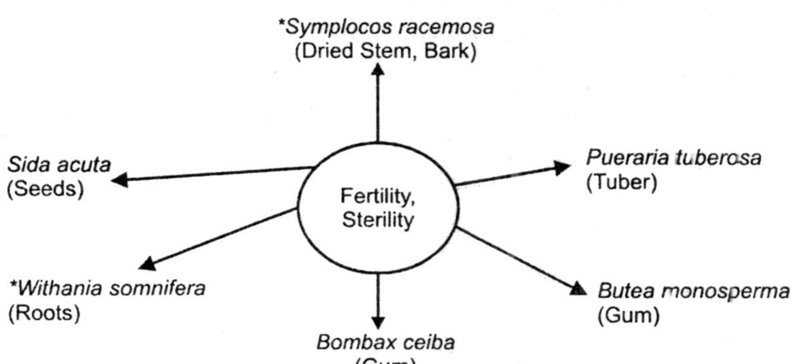
Fertility, Sterility
*Symplocos racemosa (Dried Stem, Bark)
Sida acuta (Seeds)
Pueraria tuberosa (Tuber)
*Withania somnifera (Roots)
Butea monosperma (Gum)
Bombax ceiba (Gum)

Medicinal Plants from Arid and Desert Areas

Thar desert of India is the seventh largest desert of the world and one of the most inhospitable eco-geographical region of India. It is spread over the states of Punjab, Haryana, Rajasthan and Gujarat. It covers 12 per cent of India's geographical area; besides 26 per cent area is semi arid. The hot desert area of Rajasthan has 17.44 million human and 23.33 million livestock population (2001 census). The average annual rainfall is only 100 mm to 500 mm, occurring eratically between July and September. Maximum area is covered by sand dunes ; the rest by gravel plains. The vegetation is very scanty. The forest cover is less than 3 per cent which is depleting fast due to increase human and livestock population. Of the total number of plant species 45 species are considered to be rare and/or either endangered. The area has a large number of plants of economic importance and medicinal use.

K.K. Chaudhuri (Ind. For. March,2005) records 157 medicinal plants belonging to 124 genera and 62 families. 27 species are monotypic which shows the vulnerability of these species.

Important cultivated species are–Bacopa monnieri, Boerhaavia diffusa, Plantago ovata, Plumbago zeylanica.

Of the total 157 medicinal plant species, 42 are cultivated, 115 are wild. 110 species face ruthless collection depleting the very sustainable existence.

Of 157 plant species, 85 are herbs, 51 shrub, 21 trees. There is only one perenninal Sedge and a climber and annual grass.

Though most of these listed plants have been included in the list of midicinal plants of India, recorded so far, the list prepared by Chaudhuri is presented with some modification as some of the properties of plants used by local people differ from recorded used. (*Source* : K.K. Chaudhari, Ind. For. March, 2005).

Table 11.2 : Medicinal Plants of Arid and Desert Areas

Scientific Name	Local Name	Parts Used	Medicinal Uses	Status
Abrus precatorius	Ratti, Chirmi	Seed	Sore throat, cough, chronic inflammation of mucus membrance of eyeball & eyelid.	S
Abutilon indicum	Kanghi, Tara-kanchi	Whole plant	Astringent, anthelminitic, diuretic alexiteric, toothache, demulcent tonic, rheumatism, piles, laxative, gonorrhoea.	S
Abutilon indicum	Kanghi	Seed	Tonic, gonorrhoea, bladder stone.	S
Acacia nilotica	Babool	Leaves, Bark gum	Eye sores in children, sexual impotency, diseases of urino-genital system. Astringent, demulcent, asthma, diarrhoea.	PS
Acacia genegal	Kumatia	Gum	Emollient, demuleent, burns, sore nipples, haemorrhage, intestinal mucous, food for diabeties	PS
Achyranthes aspera	Andhi-jaro, Apamarg	Whole plant	Asthma, Cough,cholera renal dropsy, gonorrhoea, eye dis. Piles, laxative, purgative anthelminitic, anti-periodic in maleria, insect bite, pneumonia hydrophobia.	PS
Adhatoda zeylanica	Adusa	Whole plants	Diuretic bronchitis gonorrhoea, jaundice, anti-spasmodic, ophtalmic, heart trouble, loss of memory, leucoderma.	S
Aegle marmelos	Bel	Root, Leaves, Fruit	Abdominal pain, heart palpitation urinary troubles, laxatives, febrifuge, ophthalmic, deafness, dysentery.	PS
Aerva persica	Bui	Flower	Information. Swelling demulcent, diuretic	PS
Ageratum conyzoides	Doachuntry	Leaves, root	Styptic, sores cuts; Root: antilitithic	PS

contd...

Table 11.2 – *contd...*

Scientific Name	Local Name	Parts Used	Medicinal Uses	Status
Alianthus excelsa	Ardu	Bark	Appetizer, arthelmintic, dysentery chronic bronchitis	PS
Alhagi pseudalhaqgi	Jawasa	Whole plant	Laxative, diuretic, expectorant, piles, rheumatism	
Alternanthera sessilis	Bhaji	Whole plant	Lactagogue, febrifuge, eye wash galactagogue.	PS
Amaranthus viridis	Jangli chauli	Leaves	Emollient, Laxative.	PS
Argemona mexicana	Satyanashi	Whole plant	Eczema, skin disorbers, blood purifier, laxative, emetic, expectorant, demulcent, jaundice, asthma diuretic, leucoderma, liniment of head, aphrodisiac, whooping cough.	PS
Aristolochia bracterolata	Kiramar, Hukka-bel	Whole plant	Anthelmintic, emmenagogue purgative, fever, joint pain, uncer, eczema.	S
Asparagus racemosus	Satawari	Root	Aphrodisiac, laxative, expectorant, galactagogue, tuberculosis, leprosy, epilepsy, night blindness, diseases of kidney, liver blood, eye and throat.	T
Azadirachata indica	Neem	Whole plant	Anthelmintic, Diuretic, blood and skin diseases, leprosy, insectididal, opthalmia, biliousness,.	PS
Bacopa monnieri	Brahmi	Whole plant	Nerve tonic, memory promoter, epilepsy, insanity.	S/T
Balanites aegyptiaca	Hingotio, Hingota	Whole plant	Purgative, anthelmintic, alexipharmic anti-dysentric, analgesic, leucoderma, ulcers, skin diseases, whooping cough, boils, leucoderma, skin.	T
Barleria cristata	Raktajhinti	Whole plant	Inflammation, fever bronchitis biliousness, tympanitis	S
Barleria prionitis	Bajra-danti, Kala bans, Chapri	Whole plant	Alexiteric, Bronchitis, blood complanists, leucoderma, laceration, cracking of feet, fever, catarrhal affection, cough, diarrhoea, toothache, boils glandular swellings, dropsy.	S

contd...

Table 11.2 – *contd...*

Scientific Name	Local Name	Parts Used	Medicinal Uses	Status
Bergia suffruticosa	Kankaio, Karbuji	Leaves	Sores, broken bones.	S
Blepharis linaraefolia	Bhangari, Bili-Khoja	Seed	Earache, tonic, increase milk production.	S
Boerhavea diffusa	Purarnava	Whole plant	Astringent, biliousness, anaemia, night blindness, leucoderma, inflammation blood purifier, diuretic, laxative, expectorant, intestinal inflammation, fever.	PS
Boswellia serrata	Salar, Salaran	Bark, fruit, gum	Biliousness, dysentery, skin diseases, ulcer, blood purifier, leucoderma, piles, antipyretic, astringent, emmenagogue, rhematism, vaginal discharge, diabetes, bronchitis.	PS
Butea monsperma	Palas, Dhak	Root, bark, leaves, flower, gum	Night blindness, elephantiasis aphrodisiac, laxative, dysentery, stomach worms, piles astringent, tonic, eye diseases, diuretic, chronic diarrhoea, round worm.	PS
Cadaba fruticosa	Dabi, Kodhab	Root, Leaves	Anthelmintic, emmenagogue, antiseptic, purgative, urinary obstruction.	S
Callogonium polygonoides	Phog, Phogda	Root Leaves	Washing eyes, sore gum.	PS
Calotropis procera	Madar	Whole plant	Toothache, stomachache, appetizer, piles, asthma, tonic.	PS
Copparis deciduas	Ker	Whole plant	Carminative, aphrodisiac, appetizer emmenagogue, alexipharmic, lumbago, rheumatism, hiccup analgesic, diaphoretic, laxative, anthelmintic, ulcer, cough, asthma, piles, cardiac troubles, inflammation.	PS

contd...

Table 11.2 – *contd...*

Scientific Name	Local Name	Parts Used	Medicinal Uses	Status
Caraluma edulis	Pimpa	whole plant	Anthelmintic, leprosy, blood diseases.	S
Cardiospermum halicacabum	Chirphuta	Root, Leaves	Diaphoretic, diuretic, gonorrhoea, rheumatism, lumbago, emmenagogue.	S
Caropegia bulbosa	Khapparkada	Tubers	Digestive tonic.	S
Cassia auriculata	Anwal, Tarawar	Whole plant	Skin diseases, astringent, anthelmintic diabetes, urinary disorders, conjunctivitis.	S
Cassia occidentalis	Anwal	Whole plant	Skin diseases, astringent, anthelmintic.	PS
Cassytha filiformis	Amebel	Whole plant	Diuretic, dysentery, ulcer, tonic, gonorrhoea, rickets, leucorrhoea.	PS
Celosia argentea	Surh Garke	Leaves, Seed	Antipyretic, aphrodisiac, liver tonic, gonorrhoea.	PS
Chenopodium album	Bathua	Whole plant	Appetizer, anthelmintic, diuretic, laxative, aphrodisiac, abdominal pain, eye disease, piles, tonic, diseases of blood, heart and spleen.	PS
Chenopodium ambrosioides		Whole plant	Carminative, emmenagogue, pectoral complaints, amenorrhoea, nervous affection.	PS
Chrozophora rottleri	Shadevi	Whole plant	Emetic, corrosive.	PS
Citrullus colocynthis	Tumba, Indrayan	Root, Fruit	Purgative, jaundice, rheumatism, urinary disease	C
Cleome gynandra	Safed bagro, Karalia	Seed. Leaves	Anthelmintic cough sores, rubefacient intermittent fever, muscular pain, rheumatism, headache, intestinal.	PS

contd...

Table 11.2 – *contd...*

Scientific Name	Local Name	Parts Used	Medicinal Uses	Status
Cleome vahliana	Madhi , Khiramar, Noli, Nodi	Seed, Leaves	Carminative, anthelmintic, rubefacient, vesicant, piles, round worms, leucoderma, skin diseases, earache, fever, dysentery, paratyphoid, bronchitis, gonorrhoea.	S
Cleome viscose	Handi-bagro, pilihulhul	Whole plant	Laxative, anthelmintic, diuretic, ulcer leprosy, malaria, piles, lumbago.	S
Clerodendrum phlomidis	Arni	Root, leaves	Laxative, alexipharmic, anaemia diabetics, chyluria, bronchitis, dyspepsia, inflammation, piles.	S
Chloria ternalea	Gokari	Root, Seed	Laxative, diuretic, alexiteric, anthelmintic, brain tonic, corneal ulcer elephantiasis, leucoderma.	S
Cocculus hisutus	Chhireta	Root, Leaves	Alexipharmic, antipyretic, laxative, soporific, venereal pain.	S
Cocculus pendulus	Poilawan	Leaves	Skin diseases.	S
Commelina benghalensis	Bukana	Whole plant	Emollient, leprosy.	PS
Commelina obiqua	Kanjuna	Root	Vertigo, laxative. biliousness, fever.	PS
Commiphora wightii	Guggal	Gum	Laxative, aphordisiac, alternative, anthelmintic, billiousness, ulcers, fistula, piles, pyorrhea, chronic, tonsillitis, pharyngitis, chronic bronchitis, chronic nasal catarrh, laryngitis, phthisis, urinary troubles, asthma, cough & cold, tumours, leucoderma, lowering blood cholestrol.	T
Convolvulus microphyllus	Santri	Whole plant	Laxative, brain tonic.	S

contd...

Table 11.2 – *contd...*

Scientific Name	Local Name	Parts Used	Medicinal Uses	Status
Corchorus depressus	Chamkas, bahuphali	Whole plant	Sexual impotency, weakness, fever, demulcent, dysentery, laxative, appetizer, anthelmintic, antiperiodic carminative, dyseopsia, catarrh, liver, disorder.	C
Corchorus olitorius	Champghas	Seed, leaves	Alternative, alexiteric, diuretic, tumours, gonorrhoea, cystitis, dysuria.	C
Corchorus tridens	Kag-nasha	Whole plant	Gonorrhoea, tonic.	
Cordia gharaf	Godela	Bark. Fruit	Astringent, headache, constipation, stomach worms piles toothache.	S
Cressa cretica	Lana	Whole plant	Aphrodisiac, stomachic, asthma diurtic, leprosy, billiousness, appetizer, tonic.	S
Crotalaria burhia	Shinyo	Whole plant	Hydrophobia, swelling.	S
Curumus prophetarum	Khat-Kachirio	Root, Fruit	Indigestion fever, purgative emetic.	S
Cuscuta hyaline	Amar bel, akash bel	Whole plant	Purgative, itch, protracted fevers, sores, chest pain.	PS
Cuscuta reflexa	Akash-bel	Whole plant	Aphrodisiac, diuretic, paralysis, heart & spleen disease, lumbago, emmenagogue, sedative, biliousness.	PS
Cynodon dactylon	Dub	Whole plant	Astringent, diuretic, dropst, cut, wound, genital urinary disorder.	PS
Cyperus rotundus	Motha, Mothee	Root-tuber	Diuretic, emmenagogue, anthelmintic, diaphoretic, astringent, stimulant, stomach disorder, bowel irritation.	PS
Dactyloctemum Sp.	Malra, Manchi. Kuri	Seed	Bellyache after chilbirth, kidney pain.	PS

contd...

Table 11.2 – *contd...*

Scientific Name	Local Name	Parts Used	Medicinal Uses	Status
Datura innoxia	Dhatura	Root, leaves, seed	Toothache, insanity, catarrhal, cerebral complaints, skin diseases, lumbago, fever, asthma, hydrophobia malarial fever.	S
Datura stramonium	Dhatura	Leaves, Seed, Fruit	Antispasmodic, narcotic, anodyne, sedative, intoxicant, carbuncles.	S
Dicoma tomentosa	Vajradanti, Choloharnacharo	Whole plant	Pyorrhoea, febrifuge, febrile attacks.	S
Digera muricata	Khanjuru	Whole plant	Astringent, laxative, biliousness, urinary discharge	S
Echinops echinatus	Unt-Kanta, Unt-Katalo	Whole plant	Alterative, diuretic, nervetonic hysteria, dyspepsia, opthalmi, cough, scrofula, seminal debility.	S
Eclipta alba	Bhangro	Whole plant	Emetic, purgative antiseptic, tonic, hepatic and spleen enlargement, jaundice, catarrh, skindiseases, hair dye.	PS
Ehretia aspera	Tombolan	Fruit	Venerealdiseases.	
Enicostema hyssopifolium	Chota-chirayata	Whole plant	Blood purifier, dropsy, rheumatism, abdominalucler, hernia, sewllings, itches, malaria.	S
Eruca sativa	Taramira	Leave, seed	Stimulant stomachic. diuretic, antiscarbtuic, aphrodisiac.	S
Euphorbia caducifolia	Thor	whole plant	Carminative, purgative, laxative, appetizer, alexipharmic, rubefacient, expectorant, cutaneous eruptions, cough, carache, bronchitis, tumour, deliriem, leucoderma, piles, spleen enlargement, anaemia, ulcers, fever.	S

contd...

Table 11.2 – *contd...*

Scientific Name	Local Name	Parts Used	Medicinal Uses	Status
Euphorbia hirta	Dudhali	Leaves	Dysentery, worms. colic, bowel complaints, cough, asthma, virmifuge, diseases of urino-genital tract, diarrhoea, leucorrhoea, menorrhagia.	PS
Evolvulus alsinoides	Shankhpushpi, Phooli	Leaves	Bronchitis, asthma, fever, diarrhoea, tonic, vermifuge, dysentery.	S
Fagonial retica	Jawasiao, Dhamso	Whole plant	Astringent febrifuge, tonic, small pox, dropsy, delirium, asthma, cough, fever, dysentery, skin diseases, abortion.	S
Fersetia hamiltonii	Hiran-chanbba	Whole plant	Rheumatism, Cooling.	S
Feronia limonia	Kaitha	Leaves, seed, Fruit	Cough, dysentery, alexiteric, heart diseases, aphrodisiac, leaucorrhoea, opthalmia, billiousness.	S
Ficus benghalensis	Bargad	Whole plant	Biliousness, ulcer, erysipelas, vaginal complaints, fevers, liver troubles, diabetes, cooling tonic.	PS
Ficus religiosa	Peepal	Whole plant	Leucorrhoea, biliousness, ulcer, diseases of vagina and uterus, alexipharmic.	PS
Furmaria indica	Pithpaparo	Whole plant	Diuretic, diaphoretic, aperients, antispasmodic CNS depressant.	S
Glinus lotoides	Hata, Badka, Matter Gandhi-butti	Whole plant	Diuretic, purgative, boils. wounds, indigestion, bilious attacks.	S
Grewia tenax	Gangeran, Gangir, Gahgi	Wood. Root	Cough, pain, diarrhoea.	S
Haloxylon recurvum	Khar	Whole plant	Ulcer.	S

contd...

Table 11.2 – *contd...*

Scientific Name	Local Name	Parts Used	Medicinal Uses	Status
Heliotropium ellipticum	Arkali	Whole plant	Laxative, diuretic.	S
Hoppea dichotoma	Ramjetta	Root	Piles, Snake bite.	S
Indigofera linifolia	Sidio bakario	Flower	Febrile eruptions, amenorrhoea.	S
Indigofera oblongifolia	Khuara	Whole plant	Root : Appetizer, rheumatism; Whole plant: spleen and liver problems.	S
Indigofera tinctoria	Neel	Whole plant	Laxative, expectorant, alexipharmic, anthelmintic, leucoderma, abdominal complaints, heart disease.	S
Ipomea headeracea	Kirpan-beli	Seed	Laxative, carminative, fever, abdominal & liver troubles, leucoderma.	S
Justica procumbens	Kagner	Whole plant	Biliousness, intoxication, fever, diuretic, enriches blood, leprosy, mental and blood diseases.	S
Lepidagathis trinervis	Unt-Katalio	Seed	Cooling drink, tonic.	S
Leucas aspera	Chota-halkusar	Leaves	Chronic rheumatism, psoriasis, skin diseases, swellings.	S
Leucas caphalotes	Vdapata	Whole plant	Diaphoretic, stimulant, laxative, bronchitis, jaundice, dyspepsia, paralysis, leucoderma, urinary discharge, fever, scorpion sting.	S
Malva parviflora	Khumbasi	Leaves, Seed	Nerve tonic, profuse menstruation, wounds & swellings; Seed: demulcent in cough, bladder uncler.	PS
Merremia tridentatea	Parasarini	Whole plant	Laxative tonic, rheumatism, piles urinary disorders.	
Mimosa himata	Shiah-Kannta, Jinjanio	Seed	Stimulant, Weakness.	S

contd...

Table 11.2 – *contd...*

Scientific Name	Local Name	Parts Used	Medicinal Uses	Status
Moringa oleifera	Sanjna	Whole plant	Aphrodisiac, alexiteric, analgesic, anthelmintic, ulcers, heart troubles, ophthalmia, muscular and spleen diseases.	PS/C
Ocimum americanum	Ban-tulsi	Whole plant	Fever, parasitical skin diseases cold, cough.	PS
Ocimum canum	Bapchi	Leaves, Seed	Fever, parasitic diseases, expectorant, anti-catarrhal, nasal haemorrhage, anti-rheumatism.	PS
Oxilis corniculata	Khatari, Khatii-buti	Whole plant	Appetizer, dysentery, diarrhoea, refrigerant, stomachic, anti-scorbutic, piles, skin diseases.	S
Oxystelma esculentum	Dudhialak	Whole plant	Diuretic, laxative, aphrodisiac, leucoderma, bronchitis, expectorant, anthelmintic, gonorrhoea.	S
Pacicum antidotale	Gramma, Garmano	Whole plant	Small pox, wound,	S
Pedalium murex	Baragokhru	Root, Leaves, Fruit	Anti-Biliousness, gonorrhoea, dysuria, anti-spasmodic, aphrodisiac, diuretic, demulcent, emmenagogue.	S
Peganum harmal	Harmal, Isband Gandhio	Whole plant	Emmenagogue, galactagogue, aphrodisiac, abortifacient, toothache; Seed: expectorant, anthelminitic, lumbago, Colic, kurinary troubles, opthalmia, rheumatism, bronchitis.	S
Pentratropis spiralis	Kauathodi	Whole plant	Whole plant: leucoderma, biliousness, piles, cough, inflammation; Root: astringent, gonorrhoea.	S
Pergularia demia	Ganderio-ki-bel, Utran, Manda-singi	Whole plant	Anthelminitic, expectorant, catarrhal affections, infantile diarrhoea, asthma, rheumatism, carbuncle, purgative, blood pressure.	S

contd...

Table 11.2 – *contd...*

Scientific Name	Local Name	Parts Used	Medicinal Uses	Status
Periploca aphylla	Barri	Whole plant	Tumkours, swellings, purgative.	S
Phaseolus triolobus	Jangli-Moth, Arak-Munhani	Leaves	Sedative, intermittent fever.	S
Phyllanthus fraternus	Kanocha, Gugaria	Whole plant	Diuretic, Dropsical affections, gonorrhoea, genito-urinary diseases, sores, stomachic, dysentery.	S
Phyllanthus maderaspatensis	Hazarmani	Leaves, Seed	Expectorant, diaphoretic, carminative, laxative, diuretic, bronchitis, earache, ophthalmia, liver tonic.	S
Phyllanthus niruri	Bhonyaabali	Whole plant	Stomachic, dysentery, dropsy, ulcer, wounds, ringworm.	S
Physalis peruviana	Baripopatan	Whole plant	Diuretic.	S
Plantago ovata	Isabgol	Seed	Astringent, tonic, biliousness, cough, dysentery, leprosy.	S
Polycarpea corymbosa	Dholphuli	Whole plant	Strangury, urinary calculi, ulcer.	S
Plumbago zeylanica	Chitrak	Root, Bark	Dysentery, leucoderma, piles, inflammation, rheumatism, bronchitis, anaemia, liver and intestinal complaints.	S
Portulaca oleracea	Lunkha	Leaves, Stem	Alexipharmic, laxative, diarrhoea, asthma, ulcer, blisters, boils, dysentery, leprosy, piles, kidney and spleen diseases, burning sensation.	S
Portulaca quadrifida	Lunki	Whole plant	Alternative, laxative, asthma, cough, urinary, discharge, ulcers. eye & skin diseases.	S
Prosopis.cineraria	Khejri	Bark, Pods	Rheumatism, astringent.	PS

contd...

Table 11.2 – *contd...*

Scientific Name	Local Name	Parts Used	Medicinal Uses	Status
Pulicaria crispa	Buhrna, Dhola-lizru	Whole plant	Headache, bruises.	S
Salvadora oleoides	Mitho-jal	Whole tree	Vesicant, purgative, cough, aphrodisiac, laxative, carminative, bronchitis, spleen enlargement, piles.	PS
Salvadora persica	Kharo-jal	Leaves, fruits	Rheumatism, scurvy, astringent, anthelmintic, diuretic, analgesic, liver tonic, carminative, deobstruent, aphorodisiac.	PS
Salvia aegyptiaca	Tukham malanga	Whole plant	Diseases of eye, diarrhoea, gonorrhoea, haemorrhage.	S
Sarcostemma acidium	Khir-Khimp, Samlata	Stem, root	Emetic, Dog-bite, snake-bite.	S
Schweinfurthia papilionacea	Sanipat	Leaves, fruits	Typhoid, bleeding nose.	S
Scoparia dulcis		Whole plant	Emollient, astringent, emetic, mennorrhagia, excessive menstruation.	S
Sesamum indicum	Til	Root, Seed, seedoil	Aphrodisiac, spleen troubles, plies bleeding , menorrhoea, diuretic, diaphoretic, coolling, hair growth, diarrhoea, lungs diseases, small-pox syphilis.	S
Sida acuta	Bal	Root, leaves	Aphrodisiac , demulcent in gonorrhoea, astringent, nervoues and urinary trouble.	S
Sida cordifolia	Bariar	Whole plant	Astringent, coolling, tonic emollient aphrodisiac, urinary troubles, blood disease, bleeding piles.	S
Sida ovata	Bal, Dhabi	Seed, root	Lumbago, tnic, cooling effect.	S
Sida rhombifolia	Bariara	Root, leaves	Aphrodisiac, tonic febrifugs, nervous and urinary troubles, strangury, heart diseases, piles, rheumatism.	S

contd...

Table 11.2 – *contd...*

Scientific Name	Local Name	Parts Used	Medicinal Uses	Status
Sisymbrium irio	Asalio, Khub Khala	Leaf, Seed	Cooling and refreshing drink.	S
Solanum indicum	Nar-kanta	Root, Fruits	Astringent, anthelmintic, digestive cardiac troubles, bronchitis, leucoderms, fever.	S
Solanum nigram	Makoi, Chirpoti	Whole plant	Diseases of eye, ear, nose, ulcer on the neck, laxative, boils, headache, aphrodisiac, alternative, diuretic, inflammation, asthma, bronchitis, fever, diarrhoea, hydrophobia, leucoderma, cathartic, liver enlargement, piles, dysentery.	S
Solanum surattense	Ringni, Kateli Bhoringi	Whole plant	Expectoranta cough, asthma, chest pain, catarrhal fever, vomiting, rheumatism, cough, toothache, burning feet.	S
Sporobolus helvolus	Kharia	Whole plant	Malaria.	S
Tamarix aphylla	Lal-jhau, Farash	Bark	Astringent, aphrodisiac, eczeme.	PS
Tecomella undulata	Rohiro, Rugtrora	Bark, Seed, young branches.	Eczema, abdominal & liver complaints, laxative, anthelmintic, abscess, ulcer, blood & eye diseases.	PS
Tephorsia purpurea	Sarphoka, Biyani, Dhamasia	Whole plant	Dyspepsia, chronic diarrhoea, enlarged liver, colic, stomach, troubles., alexipharmic, uncers, wounds, spleen complaints, anthelmintic, antipyretic, leprosy, asthma, bronchitis gonorrhoes, lung troubles.	PS
Tinospora cordifolia	Amrita gulbel	Root, Fruit	Emetic, visceral obstruction, rheumatism,jaundice	T
Tribulus terrestris	Gokhru, Kanti	Whole plant	Aphrodisiac, alternative, diuretic, emmenagogue, purgative, bronchitis, asthma, dropsy, rheumatism, skin diseases, leprosy.	T

contd...

Table 11.2 – *contd...*

Scientific Name	Local Name	Parts Used	Medicinal Uses	Status
Vallisneria spiralis	Sawala, Syala	Whole plant	Diuretic, anthelmintic, nephritic troubles, haemorrhage in kidney & uterus, jaundice.	S
Vernonia cinera	Sandri, Sahadevi	Whole plant	Diaphoretic, strangury, promotes perspiration, spasm of bladder, conjunctivitis, anthelmintic, alexipharmic.	S
Viola cinerea Boiss	Banafsha	Whole plant	Fever.	S
Withania coagulans	Paneer-bandh	Fruit, seed	Asthma, biliousness, strangury, emmenagogue, diuretic, ophthalmia, lumbago, liver troubles, piles.	T
Withania somnifera	Asgandha, Chirpotan	Root, tubes, seed leaves	Alternative, aphrodisiac, abortifacient, tonic, narcotic, bronchitis, psoriasis, ulcers, senile debility, rheumatism.	T
Wrightia tinctoria	Kerni	Whole plant	Aphrodisiac, biliousness, piles, skin, diseases, dropsy, aphrodisiac, anthelmintic, amoebic, dysentery.	S
Ziziphus mauritiana	Bor,Bordi	Root, Bark, fruit	Ulcer, wounds fever, diarrhoea, digestive, blood purifier.	S
Ziziphus nummularia	Jari-bar	Leaves, fruit	Cough, cold, skin diseases, astringent cooling, biliousness, sores, ulcerated gums.	S
Zygophyllum simplex	Lunwo, Alathi	Leaves	Opthalmia, leucoderma, anthelmintic.	S
Arisaema griffithii	L. Tangchit; N. Larua	Rhizome	Extract of rhizome used to relieve pain during urination	T
Arisaema nepenthoides	L. Panyal	Rhizome	Flour of rhizome used to make rotis.	T
Rapidophora decursiva	L. Tafu	Stem, Leaf	Boiled stem, cut into pieces and dried used in preparing chi. Leaves used as fodder.	S

contd...

Table 11.2 – *contd...*

Scientific Name	Local Name	Parts Used	Medicinal Uses	Status
Schizostachyum munrol	L. Payonh; N. Gopae	Culums	Long culums used for flutes, for making rope as phip (hollow straw for drinking chi) and to make poalong baskets. Considered sacred.	S
Thysanolaena maxima	L. Paashyuam; N.Kucho, Umlisho		Used locally for making brooms and also collected for market sale.	S
Yushania maling	L. Phyun; Malingo	Clums	Slender bamboo culms used to lay roofs and for basketry.	S
Cyathea spinulosa	L. Pula		Used to make planter-pots for culturing orchids and other green house- plants.	T
Angiopteris evecta	L. Pakjig, N. Gaaykhure	Rhizome, Stem	Dried rhizome and stem made into flour that is then used to make rotis and in preparing chi.	T

Table 11.3 : Medicinal plants of Sundarbans Mangrove

Sundarbans	Part's used	Medicinal uses	Status
Acanthus sp.	Fruit, Seed	The crushed fruit make a good blood purifier and dressing for boils and snake bite. The seeds have anthelmintic properties. Useful in asthma and dysopepsia.	PS
Ammania baccifera	Entire plant	Used as purgative.	PS
Avicennia sp.	Seeds, Resin	Seeds made into paste to relieve small pox ulceration. Resinous exude used for birth control purpose.	PS
Bruguiera eriopelata	Fruit	Lotion from fruits used for eye. Fruit is chewed as betel nut. Young radical used as vegetable.	S
Caesalpinia nuga	Roots	Roots diuretic, used in the treatment of stone.	R
Cerbera odollum	Bark, Nut	Bark and nuts as purgative and narcotic.	S
Cerbera sp.	Fruit	Fruit when rubbed give relief from pain of rheumatism. The sap has purgative property. Sap when extremely applied always the poisonous effect of fish stings.	S
Ceriops. sp.	Bark, Seed	Obstetric and haemorrhage cases are treated with an infusion of Ceriops bark. Seeds Yield, edible oil and fat, but not commercially used.	PS
Ceriops tagal	Root	Roots used as substitute of quinine.	PS
Derris indica	Seed	Seed powder used for bronchitis and whooping cough.	PS
Derris sp.	Stem, Root	Stems and root contian poisonous chemical to kill fish.	PS
D. triplicata	Entire plant	Entire plant used as antispasmodic and stimulant.	PS

contd...

Table 11.3 – *contd...*

Sundarbans	Part's used	Medicinal uses	Status
Excoecaria sp.	Heart wood	Heart wood and pneumatophores give scent, but not commercially used.	PS
Hibiscus tiliaceous	Leaf, Root	Decoction from leaves useful as hair restorers, expectorant and for treatment of obstinate cases of urine. Roots used in preparation of embrocations.	S
Ipomoea pes-carpae	Leaf	Leaves used for rheumatism and as astringent.	S
Kandelia sp.	Bark	Bark forms an ingredient in a mixture given for diabetes.	
Lumnitzera sp.	Bark, Root	An infusion of the bark of R. mucronata is given for haematuria. Stilt roots some times used as anchor. Root decoction used in blood pressure.	S
Sonneratia sp.	Fruit	Fruit made into poultices for sprain and the fermented juice is used to check haemorrhage. Fruits are edible.	S
	Gall & Twig	Galls and twigs used as astringent and	
T. gallica	–	for dysentery.	S
Tamarix dioica	Bark	Used as tonic and skin diseases.	S
Thespesia lampus	Root, Fruit	Roots and fruits used for gonorrhoea and syphilis.	S
Thespesia sp.	Seed, Leaf	An oinment made from seeds kill lice. The leaves furnish a specific active principle for relieving headache.	S
Trianthema portulacastrum	Entire plant	For heart disease and anaemia.	S
Xylocarpus mekongensis	Bark	Used for dysentery diarrhoea as febrifuge.	S

contd...

Table 11.3 – *contd...*

Sundarbans	Local Name	Part's Used	Medicinal uses	Status
Sikkim				
Aconitum bisma	N. Bikhma	Root	Extracted root taken against malaria	R
Mahonia nepaulenswis	L. Termokung, Kherum	Wood	Yellow wood used as Dhoop during rituals	S
Drymaia cordata	N. Abhijal	Leaf	Mixed leaves with Fragaria indica taken against jaundice. 2-3 doses claimed to be sufficient cure .	PS
Hibiscus-rosa sinensis	N. Javakusum	Flower	Paste of red flowers taken orally to counteract excessive menstrual bleeding.	PS
Prunus cerasoides	K. Kamki, N. Payung	Fruit, Leaf	Ripe fruits edible. Leaves used as fodder	PS
Rubus ellipticus	L. Kisimpat, N.Aniselo	Root, Fruits	Root paste taken orally against dry cough. Diluted paste of tender stems taken against stomach ache. Fruits edible.	PS
Rubus folliosus	L. Chhelum	Fruits	Fruits edible	S
Rubus nepalensis	L. Salum	Fruits	Fruits edible	PS
Astilbe rivularis	N. Burokhati	Leaf	Paste of leaves taken orally against menstrual dysfunction (backaches & excessive bleeding)	S
Bergenia ciliata	L. Tasjiprok, N. Pakhanbed	Root	Extracted root taken orally against food-poisoning & to relieve labour pains	S
Dichroa febrifuga	L. Gebokhanak;	Root	Extracted root taken orally against dysentery	PS
Tnchosanthes tricuspidata	L. Khampthoro; N. Indrani	Fruits	Fruits edible and used in pickles	PS

contd...

Table 11.3 – *contd...*

Sundarbans	Local Name	Part's Used	Medicinal uses	Status
Nardostachys grandiflora	L. Panpu, N. Jatamasi	Root	Powdered root taken with water against epilepsy. Roots burnt in ghee as Dhoop	T
Saussurea gossypiphora	L. Yackhhephabo	Inflore-scence	Paste of dried inflorescence applied on burns	T
Lyonia ovalifolia	L. Taksolnak; N. Angen	Leaf	Juice of leaves applied externally on skin-diseases of the rainy season. Poisonous to cattle.	PS
Diploknema butyracee	L. Nyol; N. Chiuri	Fruit	Fruits edible	S
Budleja paniculata	L. Pandamkung; N. Bhimsenpati	Leaf	Tablets of leaves mixed with riceflouract as fermentine agent giving chia bitter taste	PS
Datura metel	L. Kujuphimyun-gman N. Dhatura	Fruits, Seed	Fruits & Seeds taken against dog-bite	S
Picrorrhiza scrophulariflora	L. Humle; N. Kutki	Root	Powdered root taken against headaches & fevers	T
Phytolacca acinosa	N. Jaringo	Leaf	Green leaves used as vegetable. Boiled extract taken orally against food-poisoning.	S
Aconogonum mulla	L. Kandemdong, N. Thotnay	Stem	Juicy stem edible. Also used in chutneys.	S
Rheum australe	L. Chhucha	Root	Roots infused as tea	T
Houttuynia cordata	N. Gandhejhar	Leaf	Juice of leaves with curd taken against dysentery	T
Daphne bholua	L. Nambongkantuh; N. Kagajejhar	Root	Juice of root taken orally against food poisoning.	S

contd...

Table 11.3 – *contd...*

Sundarbans	Local Name	Part's Used	Medicinal uses	Status
Edgeworthia gardenari	L. Kuntkung; A. Argali	Bark	Bark used in making rope and for making paper.	S
Litsea cubeba	N. Siltimur	Leaf	Oil extracted from macerated leaf relieves skin itch.	PS
Viscum articulatum	L. Singthut; N. Kabra	Plant	Thick paste of plant used as cast for bone-setting. Powdered plant mixed with rice flour and taken orally during delivery lessons pain & bleeding .	S
Elatostema platphylum	L. Kanchelbhi	Leaf	Extract of boiled leaves taken for deworming. Leaves also used as vegetable.	PS
Girardinia diversifolia	L. Kujusrung; N. Bhangre sinsu	Leaf, Stem	Boiled leaves used as vegetable. Fibra from stem used to make gunny like cloth for traditional garments.	PS
Ficus glaberrima	L. Ringjikung	Leaf	Tender leaves used as vegetable and in pickle.	S
Ficus lacor	L. Sikungkung; N. Kanakpa	Leaf	Tender leavels wet-grount to chutney. Leaves used as fodder.	S
Juglans regia	L. Kalkung; N. Okhar	Bark	Paste of bark used as leech-repellent.	S
Dactylorhiza hatagirea	L. Ambulkapa; N. Panchamle	Root	Paste of root applied to heal fractures. Root brunt as dhoop.	T
Paphiopedilum venustum	L. Barif	Leaf	Juice of crushed green leaves applied on snakebite.	T
Amomum subulatum	N. Ainlaichi	Seed, Flower	Ripe seeds taken against gastric problems. Flowers used as vegetable. Outer sheath of flower used in making deep red dye.	PS

contd...

Table 11.3 – *contd...*

Sundarbans	Local Name	Part's Used	Medicinal uses	Status
Costus speciosus	L. Kafer	Leaf	Burnt leaves mixed with water and made into pellets, which are then taken orally with tea or chi to increase fertility and for conception.	S
Dioscorea glabra	L. Kiew; N. Bantarul	Root	Rootstock used as vegetable.	T
Campylandra aurantiaca	L. Barfi; N. Nakima	Inflore-scence	Inflorescence used as vegetable.	T
Calamus erectus	N. Bent-ko-tuso	Stem	Inner pith of stem used as vegetable.	T

Table 11.4 : Medicinal Plants of Bankura District (Alluvial and Lateritic Area)

Scientific Name	Local Name	Part's Used	Medicinal Uses	Status
Abutilon indicum	Potari	Leaves	Inflammation and wounds.	S
Acacia nilotica	Babla	Bark, Gum	Diarrohea, Dysentery and Diabetes.	PS
Acalypha indica	Mutka jhuri	Entire plant	Bronchites, Asthma and Piles.	PS
Achyranthus aspera	Apang	Entire plant	Purgative, Piles, Boils, Colic pain.	PS
Adhatoda vasica	Basak	Leaves	Cough, Bronchites, asthma, tonic.	S
Aerva lanata	Chaldhoa	Entire plant	Anthelmentic, Diuretic.	PS
Aganosma Caryophllata	Malati	Leaves & flowers	Leaves in biliousness, glowers in eye.	S
Ageratum conyzides	Uchunti	Leaves	Cuts and sores.	PS
Alternanthera sessiles	Sinche	Young shoot	Nutritious, febrifuge.	PS
Amaranthus spinousus	Kata-notey	Leaves & root	Boils and abcess.	PS
Ambroma augusta	Ulatkambal	Root-bark	Female diseases.	S
Andrographis paniculate	Kalmegh	Leaves & roots	Tonic, dysentery and dyspepsia.	T
Aneilema nufiflorum	Kenduli	Entire plant	Leprosy.	S
Anisomeles ovata	Gobura	Entire plant	Tonic.	S
Anona squamosa	Ata	Leaves	Insecticide & Lice remover.	C
Agremone maxicana	Shial kanta	Seeds & roots	Skin disease, seeds-laxative.	S
Aristolochia indica	Ishwarmul	Root & leaves	Tonic, Leucoderma, skin disease.	T
Asparagus racemosus	Satamuli	Root	Refrigerant, dysentery, diuretic.	T

contd...

Table 11.4 – *contd...*

Scientific Name	Local Name	Part's Used	Medicinal Uses	Status
Barleria prionitis	Kantajanti	Toot & leaves swellings	Tooth-ache, Boils, glanular	S
Bidens biternata		Root	Amoebic dysentery	S
Biophytum sensitivum	Naranga	Leaves, Seeds	Laxative	S
Boerhaavia diffusa	Punarnaba	Entire plant	Anaemia, Jaundice	S
Buttneria herbacea	Kamraj	Seeds	Laxative	S
Caesalpinia bonduc	Nata	Seeds	Malaria	S
Cajanus cajan	Arhar	Leaves, Seeds	Jaundice	C
Calotropis gigantea	Akanda	Leaves	Tonsilites	S
Capparis zeylanica	Kelekara	Leaves, Fruits	Eczema, boils, piles	S
Cassia occidentalis	Kalkasunde	Leaves	Tonic, rat-bite	PS
Catharanthus roseus	Nayantara	Leaves	Diabetes	C/T
Centratherum anthelminticum	Somraj	Seeds	Anthelmintic,, digestive	C
Chenopodium album	Behtosak	Shoots, Leaves	Laxative, anthelmintic	S
Cinnamomum tamala	Tajpat	Leaves	Stimulant, carminative, throat	S
Clerodandrum viscosum	Ghetu	Leaves & root	Skin disease, tumour	S
Clitoria ternatea	Aparajita	Seeds & root	Purgative, diuretic	C
Commelina benghalensis	Kansira	Plant & latex	Laxative, leprosy, ear-ache	S
Commelina paludosa	Jata Kansira	Plant	Skin disease	S

contd...

Table 11.4 – *contd...*

Scientific Name	Local Name	Part's Used	Medicinal Uses	Status
Corchorus olitorius	Mithapat	Plant	Tonic, diuretic	C
Coriandrum sativum	Dhane	Leaves & Frutis	Stimulant, carminative, Tonic	C
Crotalaria juncea	Atashi	Seeds	Blood purifier	C
Croton bonplandianum	Bantulsi	Leaves	Blood coagulation, antiseptic	S
Curculigo orchioides	Talmuli	Roots	Diabetes, jaundice	T
Cynodon doctylon	Durba	Entire plant	Cuts & wounds, hysteria & Epilepsy	PS
Daemia extensa	Dudhilata	Leaves	Expectorant, rheumatism carbuncle	S
Desmodium polycarpum	Bancharal	Entire plant	Tonic, cold and cough	S
Dioscorea bulbifera	Banalu	Rhizome	Dysentery, piles & ulcers	T
Eclipta alba	Keṣute	Leaves	Tonic, jaundice, hair dye	S
Elephantopus scaber	Hastipada	Roots	Diabetes	S
Emilia sonchifolia	Sudhimudi	Plant	Febrifuge	S
Enhydra fluctuans	Hingche	leaves	Laxative, antibiolous, skin & nerve	PS
Eugenia jambolana	Kalo-jam	Leaves & Fruits	Dysentery, carminative	C
Euphorbia hirta	Barakarni	Latex	Dysentery, cough, asthma, childrens worm	S
Euphorbia tirucalla	Sijlanka	Milky juice	Rheumatism & burn	C
Ficus bengalensis	Bat	Aerial root scorpion bite	Purgative, leprosy, piles.	PS
Glycosmis pentaphylla	Ashshora	Wood	Snake-bite	S

contd...

Table 11.4 – *contd...*

Scientific Name	Local Name	Part's Used	Medicinal Uses	Status
Gnaphalium luteo-album		Leaves	Astringent	S
Helicteres isora	Atmora	Fruit	Flatulence of Children	S
Heliotropium indicum	Hatisur	Leaves	Boils & wounds	S
Hemidesmus indicus	Anantamul	Root	Tonic, Scorpion-bite	T
Holarrhena antidysentrica	Kurchi	Bark & seed	Dysentery, fever, intestinal worms	PS
Hydrophila spinosa	Kulekhara	Entire plant	Jaundice, rheumatism, Urine-genital diseases	PS
Ipomoea aquatica	Kalmisak	Entire plant	Nervous & general debility	PS
Ipomoea pes-tigridis	Langulidlata	Root	Pargative, Boils & curbuncles	PS
Ipomoea quamoclit	Tarulata	Leaves	Carbuncle & bleeding piles	PS
Jatropha gossypifolia	Lal-bheranda	Shoot & Leaves	Boils, carbuncles, Dental disease	PS
Justicia gendarusa	Jagat-madan	Leaves & shoot	Rheumatism, ear-ache	PS
Lawsonia inermis	Mehindi	Leaves & bark	Skin disease, leprosy	S
Leonotis nepetaefolia	Hejurchei	Leaves & flowers	Tonic, ring warm & skin disease	S
Leonurus sibiricus	Raktadrone	Entire plant	Febrifuge	S
Leucas aspera	Shetadrone	Leaves	Scabies, insecticide	S
Lindenbergia indica	Besanti	Shoot	Bronchites	S
Martynia diandra	Bag-noch	Leaves & Fruits	Tuberculosis glands, epilepsy	S
Mirabilis jalapa	Sandhyamoni	Leaves & root	Boils, inflammation ,purgative	S
Murraya Koenigii	Curry leaf	Leaves	Tonic, Stomachic dysentery	S
Nerium odorum	Karabi (White)	Root	Skin-disease	C

contd...

Table 11.4 – *contd...*

Scientific Name	Local Name	Part's Used	Medicinal Uses	Status
Nymphoides cristata	Panseuli	Leaves	Boils, fever	PS
Ocimum canum	Bantulsi	Leaves	Parastical skin-diesese	S
Oldenlandia corymbosa	Khetapara	Entir plant	Jaundice & diseases of liver	S
Oxalis corniculata	Amrul	Entire plant	Cooling, stomachic, scurvy	S
Passiflora foetida	Jhumkalata	Leaves	Asthma, Headache	S
Phaseolus radiatus	Maskalai	Seeds	Paralysis, rheumatism, cough, fever	C
Physalis minima	Banterpari	Leaves	Ear-ache	S
Plumbago zeylanica	Chitamul	Root	Apetizer, skin-disease	S
Polycarpon loeflingeae	Gima sak	Leaves	Expectorant	S
Portulaca oleracea	Bara-nonoya	Plant	Scruvy & Liver disease	S
Pterocarpus marsurpium	Pitsal	Gums & leaves	Diarrhoea, boils, sores	PS
Ricinus communis	Rehri	Leaves & root	Leaves in ear-ache, root in gonorrhoea	PS
Santalum album	Chandan	Heart wood	Inflammation, headache	T
Scoparia dulcis	Ban-dhoney	Plant	Emetic	S
Sesbania sesban	Jayanti	Leaves	Pain of bones & joints, athelmintic	C
Sida cordifolia	Berela	Leaves	Cut and bruises	S
Sida rhombifolia	Lal-berela	Leaves & root	Swelling, rheumatism	S
Smilax zeylanica	Kumarika	Roots	Veneral diseases, rheumatism	T
Solanum surattense	Kantikari	Roots & fruits	Expectorant, Asthma, sorethroat	S
Solanum sisymbriifolium	Shet-Kantikari	Leaves	Pain, abnormal menstrual cycle	S

contd...

Table 11.4 – *contd...*

Scientific Name	Local Name	Part's Used	Medicinal Uses	Status
Solanum nigrum	Kakmachi	Entire plant	Liver disease, piles, dysentery & skin disease	S
Solanum ferox	Rambegum	Roots & fruit	Cough, asthma, fever expectorant	S
Spermacoce hispida		Plant	Stimulant, tooth-ache	T
Strychnos nux-vomica	Kuchila	Seeds	Poisonous, used in homeopathy	T
Symplocos recemosa	Lodh	Bark	Cooling, astringent, menorrhagea, Ulcer, eye disease	C
Tagetes erecta	Genda	Leaves	Piles, boils, carbunles.	C
Tamarindus indica	Tentul	Fruit	Refrigerant, digestive, carminative	C
Tephrosia purpurea	Bannil	Plant	Tonic, anthelmintic	S
Terminalia chebula	Haritaki	Fruit	Astringent, Laxative carious teeth, bleeding of gums.	C
Thevetia nerifolia	Kalkephul	Seeds	Highly poisonous	C
Tinospora cordifolia	Gulancha	Seeds & root	Chronic diarrhoea & dysentery	T
Trachelospermum fragrans	Bansful	Plant	Fever & Dysentery	S
Tragia involucrata	Bichati	Root & Fruits	Pains, external application in leprosy & boldness	S
Traint[illegible] monogyna	Set-punne	Plant	Diuretic, oedoma & dropsy	S
Tribulus terrestris	Gokhru	Fruits	Tonic & kidney diseases	T
Tridax procumbens	Tridaksha	Leaves	Haemostatic	S
Vitex negundo	Buan	Leaves	Tonic, acute rhumatism & swelling	PS
Waltheria indica	Khardudhi	Plant	Cold & cough	C
Ziziphus mouritiana	Kul	Leaves, Fruit & root	Dysentery	PS

contd...

Table 11.4 – *contd...*

Scientific Name	Local Name	Part's Used	Medicinal Uses	Status
Medicinal and Ethnobotanical plants to mark the value of Tripura forests:				
Abelmoschus esculentus	Bhindi	Capsule	Urinary and skin diseases	S
Abrus precatorius	Kunch	Seed	Purgative and Lonnie	T
Abutilon indicum	Patter	Whole plant	Expel worm	S
Acalypha indica	Muklajhu	Leaf	Nasal and wounds	S
Achyranthes aspera	Apang	Plant, leaf, root, seed	Wounds, cough	S
Aegle marmelos	Bel	Root and leaf	Rabies, heart, diarrhoea, dysentery	C
Agave americana		Stem and leaf	Toothache	C
Ageratum conyzoides	Pichas	Leaf	Eye, wounds, fever eczema	S
Albizzia lebbek		Leaf and bark	Blood, skin, piles	PS
A. procera		Whole plant	Skin	PS
Ammania baccifera		Whole plant	Fever and child diseases	S
Andrographis paniculata	Kalmegh	Stem and leaf	Stomachache, fever ulcer, skin desease	T
Annona squamosa	Sitaphal	Bark, root, leaf and fruit	Diarrhoea, Dysentery, Hysteria	C
Argemone maxicana	Satavnasi	Juice, root and seed	Skin eye, and expel worm	S
Asclepias curassavica	Dhudi	Latex and root, tuber, root, leaf	Leucoerma and asthma, stomach ache, piles kidney, liver, urinary, fever.	S
Azadirachta indica	Neem	All plant parts	Toothache, skin, eye, diabetes, fever, insecticide.	PS

contd...

Table 11.4 – *contd...*

Scientific Name	Local Name	Part's Used	Medicinal Uses	Status
Baliospermum montanum	Danti	Root and seed	Pain, piles, wounds, jaundice, purgative	T
Bauhinia purpurea		Root, leaf bark, flower	Fever, headache, diarrhoea, rheumatism, curative	PS
Bauhinia racemosa		Leaf	Diarrhoea, dysentery	PS
Bauhinia variegata		Root, bark, flower	Skin, diarrhoea, worms, tuberculosis	PS
Biophytum sensitivum	Lajbanti	Plant & seed	Stomachache	S
Boerhaavia diffusa	Punarnava	Root and leaf	Female disorders, lever, blood and heart	S
Bombax ceiba	Semul	Resin, gum and flower	Diarrhoea, female disorders	PS
Butea monosperma	Palas	Root, bark, leaf, flower and seed	Eye, blood, diarrhoea, dysentery, piles, worm	PS
Caesalpinia bonduc		Bark, leaf, seed, seedoil	Fever, toothache, diarrhoea, ear and bleeding	S
Calotropis gigantea		Root, juice, root and leaf	Toothache, antidote, asthma, cough	S
Cardiospermum halicacabum		Whole plant	Rheumatism	S
Careya arborea	Kumbhi	Bark, dried calyx, leaf	Stomachache, diarrhoea, eye	PS
Cassia fistula	Amaltas	Leaf, fruit, seed	Wormicide, skin, toothache, fever	PS
Cassia tora		Root bark	Antidote, skin, wormicide, cough	PS

contd...

Table 11.4 – *contd...*

Scientific Name	Local Name	Part's Used	Medicinal Uses	Status
Centella asiatica		Whole plant, leaf	Leprosy, cholera, boil, cough	S
Cissampelos pareira		Root and leaf	Skin, wounds, diarrhoea, sore	S
Cissus quadrangularis	Harjori	Whote plant stem	Scurvy, asthma, bone fracture, syomachache.	C
Cleome gynandra		Root, bark root	Wounds, headache, cough, cholera	S
Clitoria ternatea		Root	Swelling, leprosy	C
Cordia dichotoma		Bark, fruit, leaf, kemel	Pain, chronic ulcer, headache	T
Curculigo orchioides		Rhizome and roots	Asthma, jaundice, piles, cuts, female disorders	T
Cucumis sativum	Khirra	Fruit and seed	Skin, piles, indigestion	C
Curcuma longa	Haldi	Rhizome	Indigestion rheumatism	C
Cuscuta reflexa		Plant	Purgative, fever, bodyache, stomach	S
Datura metal	Kaladhatura	Leaf, fruit	Antiodote, skin, headache	S
D. stramonium	Datura	Leaf, seed	Asthma, teeth, mouth, skin	S
Dioscorea bulbifera		Tuber	Ulcer, piles, dysentery, constipation	T
Diospyros montana	Bislendu	Leaf and Fruit	Fish poison, swelling	S
Eclipta prostrata		Leaf	Jaundice, headache, fever	S
Elephantopus scaber		Plant and root	Antidote, headache, urinary	S
Emblica officinalis		Leaf, fruit, seed	Purgative, hair growth, eyes, dysentery	C/PS
Euphorbia hirta	Dodhi	Plant, latex, leaf	Asthma, disorders of womem, eye, burns, pain	S
Ficus bengalensis		Root, bark, fruit	Gonorrhoea, liver, dysentery, diabetes piles, toothache, nasal	PS

contd...

Table 11.4 – *contd...*

Scientific Name	Local Name	Part's Used	Medicinal Uses	Status
F. hispida		Fruit	Asthma, constipation	PS
F. religiosa		Bark juice	Bone fracture, asthma	PS
Flacourtia indica		Leaf, fruit	Toothaches, disorder of women	S
Flemingia microphyla		Root	Ulcer, swelling, leprosy	S
Gardenia latifolia		Leaf	Wounds of cattle	S
Garuga pinnata		Stem, leaf	Eye, asthma	PS
Gmelina arborea		Leaf, root bark	Worm, gonorrhoea	C
Helicteres isora		Leaf, fruit	Stomachache, diabetes, dysentery	S
Hemidesmus indicus	Anantamul	Whole plant	Asthma, tonsil, stomachache	T
Holarrhena pubescens		Bark	Diarrhoea, syomachache, dysentery, indigestion	PS
Holoptelea integrifolia		Leaf	Boild	S
Hyptis suaveolens	Bantulsi	Whole plant, leaf	Skin, Stomachache, cold	S
Ichnocarpus frutescens		Whole plant	Urinary, fever, skin, fractures	S
Ipomoea aquatica	Kalmi	Plant, flower	Women's disorder, eye	PS
Justicia adhatoda	Bhasake	Whole plant, leaf	Toothache, diarrhoea, dysentery, cough, asthma	S
Lagerstroemia parviflora		Bark, Leaf and seed	Purgative, astringent	PS
Lowsonia inermis	Mendhi	Root and leaf	Jaundice, hairgrowth	C
Limonia acidsima	Kaith	Stem bark	Skin and antidote	S

contd...

Table 11.4 – *contd...*

Scientific Name	Local Name	Part's Used	Medicinal Uses	Status
Linum usitatissimum	Alsi	Seed, mucilage	Urinary, rheumatism, cough and cold	S
Mallotus phlippinensis	Rohini	Fruit, seed	Pain, skin, cuts, wouds	PS
Mangifera indica	Am	Karnal, fruit, seed	Urinary, purgative, diarrhoea, nasal	C
Melochia corchorifolia	Bilpal	Leaf	Dysentery	S
Mimosa pudica		Root, leaf	Dysentery, piles	S
Mimusops elengi		Flowers	Fever, headache, pain	C
Mtragyna parviflora		Bark	Fever, pain, diarrhoea	S
Momordica charantia	Kerala	Leaf, flower	Skin and diabetes	C
M. dioica	Sangli, kerala	Root	Skin, piles, liver	C
Moringia olifera		Leaf, flower, seed	Skin, pyorrhea and pain	C
Mucuna pruriens		Root, pod	Dysentery, wormicide	S
Murraya koenigii		Sed and leaf	Cuts, diarrhoea	C
Nelumbo nucifera		Rooft, flower, seed	Diarrhoea, dysentey, cholera	T
Ocinum sanctum	Tulsi	Root, leaf	Antidote, fever, cough, cold, earache, skin	C
Parthenium hysterophorus		Plant, root	Tonic, dysentery, skin	S
Pongamia pinnata	Karank	Seed and oil	Skin	PS
Portulaca oleracea		Skin	Burns	S
Pterospermum acerifolium		Leaf	Wounds	S

contd...

Table 11.4 – *contd...*

Scientific Name	Local Name	Part's Used	Medicinal Uses	Status
Raphanus sativus		Root	Stomachache, skin, heart, cholera	C
Sida cordifolia		Root, Stem, leaf	Cuts, boil, gonorrhoea	S
S. rhombifolia		Root, leaf	Fever, heart, burns, piles	S
Semecarpus anacardium		Bark, seed, fruit	Liver, skin, pain, antidote	C
Solanum migran		Leaf, fruit	Dysentery, skin, eye, kidney, heart, liver, cold, cough, eye, asthma, skin, pyrrhoea	S
Syzygium cuminii	Jamun	Bark, fruit, seed	Diarrhoea, urinary, diabetes	C
Tamarindus indica	Imli	Leaf and seed	Eye, boil	C
Tectona grandis		Wood, flower, seed	Pain, headache, eye	PS
Tephrosia purpurea	Sorphonka	Whole plant, root	Asthma, piles, warts, weakness	S
Terminalia arjuna	Arjun	Bark, leaf	Aulidote, pain, monk, earache	PS
Terminalia bellerica	Bohera	Fruit	Stomachache, rheumatism, piles	PS
T. chebula	Horra	Fruit	Purgalue, asthma, cough	C
Tinospora cordifolia	Gurbel	Whole plant	Tonic, eye, fever	T
Tridax produmbens		Whole plant, leaf	Bleeding, cold, diarrhoea, dysentery	S
Tricumfetta rotundifolia		Seed, oil	Skin	S
Vernonia cinera		Root, seed	Fever, skin, insecticide	S
Vitex negundo		Branches, leaf	Fever, headache, ear wounds, swelling, asthma	PS
Woodfordia fruticosa		Leaf, flower	Dysentery, cough, skin eye.	PS

Table 11.5 : Medicinal Plants of Buxa Tiger Reserve

Species	Part's Used	Medicinal Uses	Status
Abrus precatorius	Seed	As a purgative and tonic	S
Abutilon indicum	Whole plant	Expel worms	S
Acacia catechu	Bark	Stomachache	PS
Acalypha indica	Leaf	Nasal and wounds	S
Achyranthes aspara	Plant, Root, Leaf and seed	Wounds, Ear, Cough Asthma, Stomachache, Disorders of women and Piles	C
Agave americana	Stem and Leaf	Toothache	S
Ageratum conyzoides	Leaf	Eye, Wounds, Eczema and Fever	PS
Albizzia lebbeck	Leaf and Bark	Blood, skin, piles, Antidole	PS
A. procera	Whole plant	Fodder skin	C
Allium cepa	Bulb	Malaria, Asthma, Ear, Eye, Menses, Child diseases and Skin.	C
A. sativum	Blue and Clove	Eye, Heart, Asthma, Ear, Paralysis, Pain and Rheumatism	S
Ammannia baccifera	Whole plant	Fever and Child diseases	T
Andrographis paniculata	Stem and Leaf	Stomachache, Fever, Skin diseases and Ulcer	S
Argemone maxicana	Juice, Root & Seed	Skin, Eye and Expel worms	S
Asclepias curassavica	Latex and Root	Leucoderma and astha	S
Asparagus recemosus	Tuber, Root, Leaf and Pulp	Stomachache, Piles, Kidney, Liver, Urinary, Fever and Disorder of women	T

contd...

Table 11.5 – *contd...*

Species	Part's Used	Medicinal Uses	Status
Azadirachta indica	All parts of plant	Toothche, Skin, Antidote, Eye, Diabetes, Urinary, Fever and insecticides	C
Baliospermum montanum	Root and seed	Pain, Skin, Piles, Wounds, Splen, Jaundice and Purgative	T
Bauhinia purpurea	Root, Leaf, Bark and Flower	Fever, headache, Diarrhoea Rheumatism and Purgative	PS
B. racemosa	Leaf	Diarrhoea and Dysentery	PS
B. variegata	Root, Bud, Bark and Flower	Skin, Diarrhoea, Worms, Tuberculosis and Wounds	PS
Bergia ammannioides	Whole plant	Bone fracture and Menstrual disorders	T
Biophytum sensitivum	Plant and seed	Stomach ache	S
Boerhavia diffusa	Root and Leaf	Disorder of womem liver, Antidote, Blood and Heart.	S
Bombax ceiba	Resin, Gum & Flower	Diarrhoea whole plant disorders women	PS
Butea monosperma	Root, Bark, Leaf Flower and Seed	Eye, Blood, Diarrhoea, Dysentery, Piles, Worms and skin diseases	PS
Caesalpinia bonduc	Bark, Leaf, Seed and Seedoil	Fever, Toothache, Diarrhoea, Ear and Bleeding.	S
C. pulcherrima	Leaf and Flower	Wounds Febrifuge	C
Calotropis gigantean	Root, Latex, Leaf and Flower	Wormicide, Fever, Cholera, antidote, Cough and Cold.	S
C. procera	Root, Juice, Rhizome and Leaf	Toothache, Antidote, Asthma and Cough	S
Cardiospermum Helicsacabum	Whole plant	Rheumatism	S
Careya arborea	Bark, Dried calyz and Leaf	Stomach ache, Diarrhoea Eye and Swelings	PS
Cassia fistula	Leaf, Fruit & Seed	Wormsicide, Skin, Toothache	PS

contd...

Table 11.5 – *contd...*

Species	Part's Used	Medicinal Uses	Status
C. tora	Root, Leaf& Seed	Fever Antidote, Cuts, Skin, Wormicide and Cough	S
Centella asistica	Whole plant and leaf	Leprosy, Brain tonic, Cholera, Boils and Cough	S
Cissampelos pareira	Root and Leaf	Skin, Wounds, Urinary, Diarrhoea, Sore and Sinuses	S
Cissus quadrangularis	Whole Plant and Stem	Scurvy, Disorder of Womem, Asthma, Wormicide and Swellings	C
Cleome gynandra	Root bark, Root	Wounds, Headache, cough, cholera and Fish poison	S
Clerodendrum indicum	Root and Leaf	Asthma, Wormicide and Swelling	S
Clitoria ten atea	Root	Welling and Leprosy	C
Cordia dichotoma	Bark, Fruit, Leaf Kemal	Pain, As a tonic, Ulcers, Headache and Wormicide	S
Curculigo orchioides	Rhizome and Root	Asthma, Jaundice, Piles, Cuts and Wounds, Disorder of women and child diseases	T
Curcuma longa	Rhozome	Indigestion, Rheumatism, Fever and Disorder of Women	S
Cuscuta reflexa	Plant	As a purgative, Fever, Body-ache and Stomach-ache	S
Dalbergia sisoo	Leaf	Eye	PS
Datura metal	Leaf, Fruit	Antidote, skin and Headache	S
D. stramonium	Leaf and Seed	Asthma, Women disorder, Teeth, Mouth and Skin	
Delonix regia	Seed gum	Phorrhoea	C
Dioscorea bulbifera	Tubers	Ulcer, Piles, Dysentery & constipation	T
Eclipta prostrata	Leaf	Jaundice, Hair growth, Headache & Fever	S
Ehretia laevis	Root, Leaf, Seed	Sex disorders and Fodder	S
Elephantopus scaber	*Plant and Root*	Antidote, Heart and Urinary	S

contd...

Table 11.5 – *contd...*

Species	Part's Used	Medicinal Uses	Status
Emblica officinais	Leaf, Fruit & Seed	Purgative, Hair growth, Eye, Scurvy, Diarrhoea, Dysentery, Antidote and as a Cool.	C/PS
Eulphorbia hirta	Plant, White juice, latex & leaf	Asthma, Disorder of women, Eye, Antidote, Burns and Pain	S
Ficus bengalensis	Root, Bark, Juice and Fruit	Gonorrhoea, Liver, Dysentery, Diabetes, Pain, Skin, Piles, toothache and Nasal.	PS
F. hispida	Fruit	Asthma and Constipation	PS
F. racemosa	Plant, Bark, Leaf, and fruit juice	Anicancer, Wounds, Piles, Diarrhoea and Dysentery	S
F. religiosa	Bark, Juice, Leaf and Fruit	Bone fracture, Antidote, Asthma, Disorder of women, Toothache	PS
Flacourtia indica	Bark Root, Fruit and seed	Skin, Dysentery, Rheumatism and speen	S
Gmelina arborea	Leaf, Root & Bark	Worm(expel), Gonorrhoea and antidote	C
Gymnema sylvestre	Root and Leaf	Stomach ache, Urinary	T
Helicteres isora	Whole plant, Root and Leaf	Asthma, Urinary, Tonsils, Stomach ache and Blood	S
Hemidesmus indicus	Wole plant, Root and Leaf	Asthma, Urinary, Tonsils, Stomach ache and Blood	T
Holarrhena pubescens	Bark and Seed	Diarrhoea, Stomach ache, Dysentery and indigestion	PS
Holoptelea integrifolia	Leaf	Boils	S
Hyptis suaveolons	Whole plant & leaf	Skin, stomach ache and cold	S
Ichnocarpus frutescens	Whole plan & Root	Urinary, Fever, Skin and Fractures	S
Ipomoea aquatica	Plant, Bud & Flower	Women disorder and Eye	PS
Justicia adhotoda	Whole plant, Leaf	Toothache, Tuberculosis, Diarrhoea, Dysentery, Cough, Asthma and Skin diseases	S

contd...

Table 11.5 – *contd...*

Species	Part's Used	Medicinal Uses	Status
Kydia calycina	Bark and Leaf	Mouth	S
Lagerstroemia parviflora	Bark, Leaf & Seed	Purgative, Astringen & to induce sleep	PS
Lannea coromandelica	Stem bark & Fruit	Cuts, Mouth, Toothache and Wounds	PS
Luffa acutangula	Leaf	Eye diseases in children	C
Mallotus philippensis	Fruit and Seed	Pain, Skin, Expel worm, Cuts, Wounds and as a Purgative	PS
Minosa pudica	Roota and Leaf	Dysentery and piles	S
Mitragyna parviflora	Bark	Fever, Pain and Diarrhoea	S
Moringa oleifera	Leaf, Flower and Seed	Skin, Urinary, Pyorrhoea and Pain	C
Mucuna pruriens	Root, Pod & Seed	Dysentery, Urinary, Wormicide	S
Murraya koenigii	Leaf	Cuts, Diarrhoea and Dysentery	C
Nelumbo nucifera	Root, Flower & Seed	Diarrhoea, Dysentery, Cholera and Eye	T
Nychanthes abhortristis	Bark, Leaf, Inflorescence, Flower, Fruit and Seed	Eye, Fracture, Fever, Hair growth, skin and cough	C
Ocimum sanctum	Root and Leaf	Antidote, Fever, Cough, Cold, Earache, Headache and Skin	C
Ougeinia oojeinensis	Bark and Wood	Fish-poison, Diarrhoea, Dysentery and Stomachache	S
Parthenium hysterophorus	Plant and Root	As a tonic, Dysentery and Skin	S
Pongamia pinnata	Seed and Oil	Skin	PS
P. oleracea	Stem	Burns (Skin)	S
Pterospermum acerifolium	Leaf	Wounds	S

contd...

Table 11.5 – *contd...*

Species	Part's Used	Medicinal Uses	Status
Ricinus communis	Root, Leaf and Seedoil	Women disorders, Pain, Jaundice	PS
Semecapus anacardium	Bark, Fruit and Seed oil	Wormicide and as a purgative, Liver, Skin, Paiss and Antidote	C
Sida cordifolia	Root, Stem & Leaf	Cuts, Urinary, Boils and Gonorrhoea	S
S. rhombifolia	Root & Leaf	Fever, Heart, Burns and Piles	S
Solanum nigrum	Leaf and Berry	Dysentery, Skin, Eye, Kidney, Heart and Lever	S
S. surattense	Root, Stem, Leaf, Fruit and Seed	Cough, Cold, Eye, Asthma, Skin and Pyorrhoea	S
Stellaria media	Bark, Fruit & Seed	Diarrhoea, Urinary and Diabetes	S
Syzygium cumini	Wood, Flower oil & seed	Pain, Headache, Eye, Skin, and Expel the worms	C/PS
Tephrosia purpurea	Whole plant & Root	Asthma, Piles, Warts and Weakness	S
Terminalia arjuna	Bark, Twigs & Leaf	Antidote, Hear, Pain, Mouth and Earache, Stomachache, Rheumatism	C/PS
T. bellirica	Fruit	Diarrhoea, Astringen, Piles, Kidney and Eye	C/PS
T. chebula	Fruit	Purgative Brain tonic, Asthma, Cough and Mouth	C/PS
Thespesia populnea	Wood Plant & Leaf	Heart, Skin and Syphilis	s
Tinospora cordifolia	Whole plant & Stone	Tonic, Eye and Fever	T
Tridax procumbens	Whole plant & Leaf	Bleeding, Cold, Diarrhoea, Dysentery, Insecticide and Wound	S
Ventilego caliculata	Bark	Skin, Urinary and as coolant	S
Vernonia cinerea	Root, Plant& Seed	Fever, Urinary, Skin and Insecticide	S

contd...

Table 11.5 – *contd...*

Species	Part's Used	Medicinal Uses	Status
Vitex negundo	Branches & Leaf	Fever, Headache, Ear, Wounds Swelling and Asthma	PS
Woodfordia fruticosa	Leaf & Flower	Dysentery, Cough, Skin and Eye	S
Wrightia tinctoria	Bark, Leaf, & Seed	Stomachache, Fever, Piles, Skin, Fishpoison and Toothache	
Xanthium indicum	Root and Fruit	Eye, Headache and Smallpox	S
Zingiber officinalis	Root or Rhizome	Cough, Stomachache, Eye Expel worms and High blood pressure	C
Ziziphus mauritiana	Root, Bark Lead, Fruit and Seed	Typhoid, Stomachache,Cuts and Wounds	C
Z. oenoplia	Root	Wounds	C

Medicinal Plants in Satpura Plateau of Madhya Pradesh

Madhya Pradesh is veritable treasure house of healing herbs which are being used in Indian System of Medicine like Ayurveda, Siddha and Unani. The plants, shrubs, roots of immense medicinal value are abundantly found in Satpura, Vindhyachal, Amarkantak, Pachmarhi and Patakot areas. Madhya Pradesh has 1,35,164 km^2 of forests which account for 30.48% of total geographical area of the state. These forests have been source of invaluable medicinal plants since the time man realized the preventive and curative properties of plants and started using them for human health care. Our old traditional Indian Systems of Medicine (ISM), one of the most ancient medicine practices known to the world, derive maximum formulations from plants and plant extracts that exist in the forests. About 400 plants are used in regular production of Ayurvedic, Unani, Siddha and tribal medicine. About 75% are from tropical and 25% from temperate forests. 30% of preparations are roots, 14% bark, 16% whole plants, 5% flowers, 10% fruits, 6% leaves, 7% seeds, 3% wood, 4% rhizomes, 6% stems, only less than 20% (including species) are cultivated (Anon., 1997).

General forest degradation processes adversely affect the resource base of medicinal plants. The rural poor, whose dependence on these products is very heavy, are the worst sufferers. The problem is compounded by market demand driven harvesting without any concern for regeneration and conservation. In the process, essential regenerative components of a plant like roots, tubers, fruits, seeds, flowers and bark are indiscriminately collected, leading to degradation and depletion and even demise of particular species. Many important medicinal plants like Rauvolphia serpentina, Curcuma caesia, Chlorophytum spp., Dioscorea spp., Gloriosa superba, Gymnema sylvestre etc. are becoming rare and some of them are critically endangered. It is estimated that 10% of all plant species are currently endangered in India.

The Satpura Plateau is a remarkable place, not only because of the large trible population and dense forests, but also because it has a lot of rare and useful natural resources including some rare species of medicinal plants, which are used for curing different kinds of diseases. Tribals and forests are symbiotically related. The tribal communities in Satpura plateau occupy forested region. They have lived in isolation but in harmony with nature. They draw their sustenance largely from the forests. Even in areas where forests do not exist, they visit distant forests periodically and try to get their traditional requirement. They have very close linkage with the forest, which they regard as their mother deity. A perusal of literature reveals that some work has been done on ethnomedicinal plants of Madhya Pradesh (Rai and Pandey,1997; Pandey and Bisaria, 1998; Pandey, 2000; Rai *et. al.*, 2000).

Herbal Garden

Some critical, endangered, vulnerable and low risk threatened species of medicinal plants of the region are Curcuma caesia, Dioscorea deltoidea, Gloriosa superba, Rauvolfia serpentina, Curculigo orchoides, Celastrus paniculatus, Baliospermum montanum, Uraginea indica, Tylophora indica, Hedychium spicatum, Gymnema sylvestre, Curcuma angustifolia and Clerodendrum serratum. A few of the medicinal plants of Satpura Plateau, which are becoming extinct, need immediate attention for conservation.

Table 11.6 : Medicinal Plants conserved in Herbal Garden in CFRHRD, Chhindwara

Scientific Name	Local Name	Part's Used	Uses and active chemical constituents	Status
Abelmoschus Moschatus	Muskdana	Seeds, roots	Seed used as cardiac tonic and aphrodisiac; source of ambrette fragrance (musk odour). It contains ambratolide.	C
Abroma augusta	Ulta Kambal	Seeds, roots and leaves	Root is abortifacient and uterine tonic. Leaf paste is used in ringworms. Roots contain choline betaine and stigmasterol.	S
Abrus precatorius	Ratti	Seeds, roots and leaves	The roots and leaves are astringent, sweet and emetic. Seeds are abortifacient and aphrodisiac, it contains glycyrrhizin and abrin.	T
Abutilon indicum	Kanghi	Roots, leaves and seed	The plants possess diuretic, demulcent and laxaative. Paste of leaves useful in boils. It contains flavours & gossypetin.	S
Acacia catechu	Khair	Bark and wood	The bark is used in conjuctivitis, stomachache and dirrhoea. It contains catechin and guym resin.	S*
Acacia concinna	Shikakai	Pods	Pods are used as hair tonic. It contains Kinmonnosides saponins.	S
Acacia nilotica	Babul	Bark, tender shoots, gum	Bark is useful in diarrhoea, dysentery, bronchitis and cough. Tender shoots are used as tooth brush. Bum constitutes galactose, arabinose, rhamnose.	S*
Achyanthes aspera	Apamarg	Roots, leaves and seeds	Expectorant, diuretic, bronchial troubles and abortifacient. Also useful in painful delivery. It contains achyranthine, betaine acdysterone and glycosides.	S

contd...

Table 11.6 - *contd...*

Scientific Name	Local Name	Part's Used	Uses and active chemical constituents	Status
Acorus calamus	Bach	Rhizome	The paste of rhizome is given for stammering in children. It contains beta-asarone, calamol, chalamene and acotamine.	T
Adhatoda vasica	Vasaka	Leaves and young twigs	Cough and bronchial troubles. Leaf decoction is used in inflammation. It contains vasicine, Vasicinol, adhatodine.	C
Aegle marmelos	Bel	Fruit and leaves	The fruit is taken in dysentery and constipation. It contains abscisic acid, marmelosin, marmin and vitamins.	C/PS
Ageratum conyzoides	Osari	Leaves	Leaves are used in wounds are sores.It contains conyzorigum, chromone and essential oil.	S
Albizzia lebbek	Kala siris	Bark, flowers and seed	Bark is astrigent, expectorant; flowers are useful in cough; seeds are used in inflammations. It contains saponin labbekanin and tannins.	S
Albizia procera	Safed siris	Bark, Seeds	Bark is astringent, expectorant; flowers are useful in cough; seeds are used in inflammation, it contains saponinprocerogenin.	S
Aloe vera	Gwar Patha	Leaves	Used in burn, eruptions, stomach troubles and acidity. It contains aloin and aloe-emodin.	T/C
Alpinia galanga	Kulanjan	Rhizome	Used as tonic deodorant and disinfectant. It yield an essential oil rich in camphor. It contains essential oil rich in cineol.	T
Amorphophallus campanulatus	Surankand	Taber	The corn is irritant, astringent, carminative in liver tonic. Tubers contain protease inhibitors, trypsin chymotrypsin.	C
Andrographis paniculata	Kalmegh	Whole plant	The herb is useful in malaria and liver discords. It contains andrographolide.	T

contd...

Table 11.6 - *contd...*

Scientific Name	Local Name	Part's Used	Uses and active chemical constituents	Status
Annona squamosa	Sitaphal	Fruits and seeds	The paste of leves is used to kill lice. Fruits is as tonic. It contains anonaine, higenamine, reticuline and squamosin.	C
Argemone mexicana	Piliktari	Roots, leaves and latex	Used in skin diseases & leprosy. Latex is useful in dropsy, jaundice, and conjunctivitis. It contains allocryptopine, berberine and maxicanic acid.	S*
Argyreia nervosa	Samudra sokh	Leaves	Used in gonorrhoea and chronic ulcers. It contains ergoline, isoergine and ergine alkaloids.	S
Artemisia maritime	Bhim Kapoor	Leaves and flowers	Flower tops are used as anthelmintic; decoction of plant is useful in fever. It yields essential oil rich in cinol.	T
Asparagus officinalis	Satavar	Roots	Roots are used as tonic. It contains satavarin, saraspogenin disogenin and glucosides.	PS
Azadirachta indica	Neem	All parts	Leaf paste is useful in skin diseases, twigs are used in toothache, seeds as insecticide. It contains azadirachtin, nimbin, mimbicidin and fatty acids.	PS
Bacopa monnieri	Brahmi	The whole plant	The whole plant is used as intellect promoting, cooling, carminative and cardiotonic. It contains baccoside, brahmine.	T
Barleria cristata	Katsaraya	Leaves and roots	Leaves and roots are used for cough and inflammations. It contains epigenin & neringenin.	T
Bauhinia purpurea	Lal Kanchnar	Leaves and roots	It is used in rheumatism, swelling, leprosy, glandular diseases and animal bites. Flowers contain astragalin, isoquercitin and quercetin.	S

contd...

Table 11.6 - *contd...*

Scientific Name	Local Name	Part's Used	Uses and active chemical constituents	Status
Bauhinia vahili	Mahul patta	Roots, leaves and seeds	Roots decoction is taken in fever. Seeds aphrodisiac. Glycosides of quercitol and kaempferol and proteins.	S
Bauhinia varigata	Kachnar	Leaves, bark and pods	The bark decoction is taken in tuberculosis. Useful in diarrhoea and worms. It contains quercetin.	S
Bixa orellana	Sinduri	Seeds	Seeds are used as astringent, gonorrhoea; leaves in snakebite. It is cultivated to obtain annotto dye (Bixin)	C*
Boerhavia diffusa	Punarnava	Root and leaves	Roots decoction is taken in jaundice, bronchitis and in kidney stone. It contains punarnavine-1&2.	S*
Bryonia lanciniosa	Shivlingi	Seeds	Seeds are taken in sterility and to get male child. Seed oil is a source punicic acid.	T
Buchnaia lanzan	Chironji	Seeds	Seeds are used as cardiac tonic. It contains protein, fat, fibre, Carbohydrates, minerals etc. Seeds yields fatty oil.	PS
Bursera panicillata	Lavender tree	Bark and wood	Wood oil is highly valued perfumes. It contains various terpenes with linalool as major constituents.	PS
Butea monosperma	Palas	Seeds, bark	Seeds are used in skindisease. Bark is aphrodisiac and anthelmintic. It contains butin, monospermoside and palasonin.	PS
Caesalpinia crista	Gataran	Seeds	Seed power is taken in stomach disorders. It contains caesalpin-F and bonduceline.	S
Callistemon lanceolatus	Bottle brush	Leaves	Used for skin ailments. Leaves yield essential oil rich in cineol.	S

contd...

Table 11.6 - *contd...*

Scientific Name	Local Name	Part's Used	Uses and active chemical constituents	Status
Calotropis procera	Aak	Flower, leaf latex	The flower are used in cough and asthma, latex is used in toothache. It contains cardinolides named gigantin, uscharidin, calotropin etc.	S
Cannabis sativa	Bhang	Leaves and seeds	It is used to treat depression, bladder inflammation and nervous disorders. It contains canabin.	C
Cardiospirmum	Kalphuti	Leaves and seeds	Used against rheumatism, arthritis and obesity. Leaves and roots are diuretic. It contains saponins.	S
Carissa carandus	Karonda	Fruits and roots	Unripe fruit-astringent, ripe; colling; root-bitter and antheminitic. Roots yield carissone, fruits ascorbic acid.	S
Carum bulbocastamum	Kala zira	Seeds	Seeds are laxative and anthelmintic. It yields a volatile oil rich in cuminaldehyde.	S
Cassia angustifolia	Seena	Leaves	Leaves are used as laxative and for intestinal troubles. It contains sennosides.	S
Cassia tora	Chakoda	Seeds and leaves	The seed paste is applied on skin diseases. It contains glucoside, amino acid, fatty, acids, emodin and sitosterol etc.	S*
Catharanthus roseus	Sadabahar	Roots and leaves	Leaf decoction in antidiabetic. Root of the plant are used to treat cancer. It contains vincristine, vinblastine, ajmalcine, vinceine and reserpine group of alkaloids.	C
Celastrus paniculatus	Mankangni	Seeds, bark and leaves	Seeds are useful in abdominal disorders and for sharpening memory and intellect. It contains malkanguniol, celapanine and paniculatdol.	T

contd...

Table 11.6 - *contd...*

Scientific Name	Local Name	Part's Used	Uses and active chemical constituents	Status
Centella asiatica	Manddok parni	Leaves	The plant is used as nervine tonic, carminative, cooling and diureic. It contains cenotic, centellic acids, Asiatic acid, brahminoside, asiaticoside.	S
Chlorophytum borivillianum	Safed mulsi	Roots	Roots are used as toinic and aphrodisiac. It contains glycosides, polysaccharides and saponins.	T
Cissus quadrangularis	Hadjori	Fleshy stem	Used to join bone fractures in animals and in human being also. It contains beta-sitosterol, amyrin, amyrone and terpernoids.	S
Citrullua colocynths	Indrayan	Leaves and fruits	Fruits are cooling, carminative, antipyretic and useful in tumours and leucoderma. It contains colocynthin, citrullol.	C
Cleome gynandra	Hur hur	Leaves and fruits	Popultice of fresh leaves are used on swellings. Flowers are used in anemia. It contains cleomin.	S
Clerodendrum seraum	Bharangi	Leaves	Leaves are useful in cough, bronchitis, intermittent fever and skin diseases. Saponin from roots, bark-catechin and luteolinc.	S
Clerodendrum phomidis	Bharangi	Leaves	It is used in fever, cough, bronchitis,and cholera. Also used as antifertility drug. Leaves contain sctellarein.	S
Clitoria ternatea	Aparajita	Roots, flowers and seeds	Root is used as diuretic, also used in snake poison. Seed and root contain tennin, seeds contain a fixed oil.	C
Commiphora mukul	Guggul	Oleo-gum resin	The olio-gum resin is used in treatment of arthritis and obesity. It contains guggulusterols-I, II, III and sterones.	T
Costus speciosus	Keokand	Rhizome	It is used to treat cough, dyspepsia, skin diseases, worms and snake bite. It contains diosgenin.	T

contd...

Table 11.6 - *contd...*

Scientific Name	Local Name	Part's Used	Uses and active chemical constituents	Status
Curculigo orchiodes	Kali Musli	Roots	Roots are used as tonic and aphrodisiac. Roots contains glycosides, polysacchride, starch, yuccagenin and sapogenin.	
Curcuma amada	Ama haldi	Tuber	Useful in sprain and brusies. Rhizome is taken in stomach pain. It yields essential oil rich in ocimene and terpene.	C
Curcuma angustifolia	Tikhur	Tuber	Rhizome is fragrant used as tonic, aphorodisiac. Useful in bronchitis, fever ans asthma. It yields cineole rich essential oil.	C
Curcuma caesia	Kali haldi	Tuber	Rhizome is useful in sprain, asthma bronchitis leucoderma. It yields camphor rich essential oil.	C
Cuscuta reflexa	Amar bel	Leaflets	Plants is purgative used in sores and fall of hairs. Plants contain cuscutalin and cuscutin.	S
Cymbopogon flexuosus	Lemon grass	Leaves	The oil is used in perfumery and synthesis of Vitamin A. Vitral is major constituents.	C/PS
Cymbopogon martini	Palamarosa grass	Leaves, inflorescence	The oil is used for pain in joints. It is rich source of geraniol.	C
Cymbopogon winterianus	Citronela grass	Leaves	The oil is used is insect repllent preparations. It is rich source of citronellal.	C
Cynodon dactylon	Dobghass	Whole plant	Useful in skin, gastro-urinary diseases. It checks bleeding from cuts and wounds. It contains tricin and flavone gltcosides.	PS
Cyperus scariosus	Gondla	Tubers	The tubers are used in diarrhoea, cough and renal troubles. It contains cyperine, aromadendrene and rotundine.	S

contd...

Table 11.6 - *contd...*

Scientific Name	Local Name	Part's Used	Uses and active chemical constituents	Status
Datura metal	Dhatura	Leaves and fruits	Leaves are used in swelling and rheumatism; flowers in asthma. It is rich source of hyoscine, scopolamine alkaloid.	T
Datura stramonium	Kala Dhatura	Leaves and fruits	Smoke of seeds inhaled in asthma. It is narcotic and antispasmodic. It contains byoscin and hyocyamine alkaloids.	T
Dioscorea daemia	Baichandi	Tubers	Tubers are used in stomach disorders, diarrhoea and piles. Fruits are used as germicidal. It contains betulin, lupeol and tannin.	T
Dioscorea daemia	Tendu	Barks, fruit	It is used in stomach disorders, diarrhoea and piles, fruits are used as germicidal. It contains betulin, lupeol and tannin.	T
Eciipta alba	Bhringaj	Leaves	Plant is used in chronic skin diseases. It is used to blacken hairs. It has stigmasterol, Egyptian and amyrin.	S
Embelia ribes	Bihdidang	Seeds	Seeds are useful in worms and abdominal disorders. It contains embelin, embolic acid.	T
Emblica officinalis	Amla	Fruits	Fruits are useful in diabetes, anemia jaundice, tridosha flatulence and grayness of hairs. It is rich source of Vitamin C	PS
Eucalyptus globulus	Nilgiri	Leaves and oil	Useful in cold and cough. Its oil is rich source of cineole.	C/PS

contd...

Table 11.6 - *contd...*

Scientific Name	Local Name	Part's Used	Uses and active chemical constituents	Status
Eucalyptus citriodora	Nilgiri	Leaves and oil	Oil is used as insect repellent and rich source of citronellal.	C/PS
Eucalyptus hybrid	Nilgiri	Leaves and oil	The oil is used in cough and cold and contains coneole, terpineiol and phellandrene.	C/PS
Evolvulus alsinoides	Shankpuspi	Whole plant	Useful in bronchitis, asthma, epilepsy, improving memory and general debility. It contains evolvine and tricontane.	S
Gloriosa superba	Kalihari	Tubers and seeds	The rhizome are useful in inflammations, labour pain and delivery of baby. It is rich source of colchicines.	C/PS
Gmelina arborea	Khamer	Bark and roots	It is astringent, bitter, digestive, cardiotonic, diuretic and laxative. It contains gmelinol and luteolin acid.	T
Gymnema sylvestre	Gudmar	Leaves	Leaves are useful in diabetes, liver disorders and jaundice. It contains gymnemic acid.	T
Hedychium spicatum	Kapoorkachri	Rhizome	Rhizome is used in bronchitis. indigestion, eye diseases and as blood purifier. Rhizome yield volatile oil.	T
Helicteris isora	Marodphalli	Root, bark and fruits	The root and bark is used in diarrhoea and dysentery. Seeds contain diosgenin.	S
Hemidesmus indicus	Anantmool	Roots	Roots are used as tonic and in skin diseases. It contains hyperoside, isoquercetin, rutin, hemidesmin 1 and 2.	T
Hibicus rosasinesis	Gurhal	Flowers, leaves	Flowers are hypoglycaemic, aphrodisiac and used in diabetes, and menstrual disorder. It contains beta-sitosterol.	C

contd...

Table 11.6 - *contd...*

Scientific Name	Local Name	Part's Used	Uses and active chemical constituents	Status
Holarrhena antidysentrica	Kutaja	Bark and seeds	Useful in amoebic dysentery, diarrhoea and asthma. It contains, conessimine, conescine, concuressine and curchisine.	PS
Hyptis suaveolens	Vantulsi	Leaves and seeds	Leaf decoction is used as eye lotion and in fever. It is used in headahe and malaria. It yields a menthol rich essential oil	S*
Iopomea digitata	Bidrikand	Roots	Roots are used as tonic aphrodisiac and purgative. Rhizomes yield taraxerol and sitosterol.	S
Ipomea nil	Kaladana	Seeds and roots	The seeds are used as anthelmintic are purgative. It contains lysergol, panniclavine.	S
Jasminum grandiflorum	Mongra	All parts	Flowers are usedful in skin diseases, headache, eye troubles and in scorpion strings. It yields fragrant oil used in perfumery.	C
Jasminimum sambac	Mongra	All parts	Whole plant imparts cooling sensation, used in insanity and weakness; useful in headache and eye troubles. It yields fragrant oil used in perfumery.	C
Jatropha curcas	Ratanjot	Seeds, oil and latex	The latex is purgative and good for wounds. The oils used in rheumatism. It contains curin, apigenin and vitaxin.	S*
Jatropha gossypifolia	Chandrjot	Seeds and oil	The latex is purgative and good for wounds. The oils used in rheumatism. It contains fatty oil. It contains jatrophone.	S*

contd...

Table 11.6 - *contd...*

Scientific Name	Local Name	Part's Used	Uses and active chemical constituents	Status
Lantana camara	Laltern	Leaves and roots	The leaves are used in ecozema. The decoction of root is used in dysentery. It contains essential oil rich in camerene.	S*
Lavandula officinalis	Lavander	Leaves and oil	It is stimulant. It provides a fragrant oil used in cosmetics. It contains linalool, linaly acetate.	C
Lawsonia inermis	Mehandi	Leaves	The paste of leaves applied on hairs to get rid of dandruff. It contains laxathones & lowsone.	C
Lepidium sativum	Chandrasur	Seeds, roots and leaves	Seeds are used in sprains and leprosy, lumbago, scurvy, asthma and cough. Plants contain glucotropoeolin.	S
Leucas aspera	Gumma	Leaves and flowers	It is used to treat cough and cold. Leaf juice is used in psoriasis, skin eruptions and swellings. It contains oleanolic acid, solic acid and fatty acids.	S
Limonia acidissima	Kaitha	Fruits	Fruits used as astringent, stomachic and stimulant. Used as remedy of insect bites. It has bergapten, marmesin, marmin.	S
Madhuca longifolia	Mahua	Leaves, flower and seeds	The bark is used for ulcers and bleeding gums. Flowers are taken as tonic. Seeds contain fatty oils and mowrin saponins.	S
Melia azedarach	Bachin	All parts	Used as insect repellent. Leaf juice anthelmintic and diuretic. Seeds yield a fatty oil. It contains bakayanin, margosine.	PS
Mentha aruensis	Japanese mint	Whole herb and oil	The oil is antiseptic, carminative and stimulant. Find useful in headache. It contains menhol as major constituent.	C

contd...

Table 11.6 - *contd...*

Scientific Name	Local Name	Part's Used	Uses and active chemical constituents	Status
Mimusops elengi	Moulsri	Bark and fruit	Bark is useful in diarrhoea and dysentery and fruits is as tonic. Seeds contain sapogenins.	C
Moringa oleifera	Sahjan	Leaves and fruits	Leaves are used in scurvy, flowers are as tonic, seeds are antipyretic, oil in rheumatism. It contains moringine, moringinine and fatty oils.	C
Mucuna pruriens	Kewanch	Seeds	Seeds are aphrodisiac and nervine tonic. It contains L-DOPA, glutathione, gallic acid, prurienidine and glycosides.	S
Murraya koenigii	Mithi neem	Leaves	Leaves are aromatic and used in preparation of curries. The bark is used in erupations. It contains koenigin and volatiles.	C
Neritum indicum	Kaner	Roots and leaves	Roots used in ulcera and decoction of leaves used to reduce swellings. It contains neriodorin, and nerioderin.	C
Nyctanthes arbortristis	Prijata	Leaves and flowers	Leaves useful in fever, rheumtism and sciatica. It contains nyctanthin.	C
Ocimum basilicum	Raam tulsi	Leaves and seeds	Carminative, diuretic, stimulant and used in gonorrhoea. It is source of eugenol and inalool.	S
Ocimum sanctum	Tulsi	Leaves and seeds	Leaves are useful in cold, cough headache and bronchial troubles and in snake bite. It contains eugenol, carvacrol etc.	C
Sida acuta	Bala	Whole herb	Herb is used to cure nervous, fever and urinary disorders. It contains ephedrine, vasicinone, vasicine and choline alkaloids.	S*

contd...

Table 11.6 - *contd...*

Scientific Name	Local Name	Part's Used	Uses and active chemical constituents	Status
Sida cordifolia	Atibala	Whole herb	Used for rheumatism, diabetes, and neurological disorders. It is used as tonic and aphrodisiac. It contains ephedrine, betaine and choline alkaloids.	S*
Sida rhombifolia	Mahabala	Whole herb	The plant is cardiotonic and useful in ulcers, skin diseases, rheumatism, sciatica and diabetes. It contains ephedrine, vasicinol and choline alkaloids.	S
Smilax perfoliata	Ramdaton	Roots	It is used in seminal emissions spermatorrhoea and weakness. Dioseginin is the major active constituents.	T
Solanum nigrum	Makoy	Fruits, leaves and seeds	It is useful in jaundice, cough, bronchitis, asthma, diarrhoea, fever ulcers and skin diseases. It contains solanine & saponins.	S
Solanum khasianum	Akarkara	Roots and flowers	The plant is used as anti inflammatory, antiarthritic and a source of steroidal hormones. It contains solalsodine.	T/C
Sphaeranthus indicus	Gorakhmundi	Fruits, leaves	The plant possesses anthelmintic, aphrodisiac and stomachic properties. It contains sphaeranthine.	S
Spilanthus calva	Bhatkatai	Roots and flowers	Useful in tooth pain and ulcers in mouth. It contains spilanthol.	S
Sterculia urens	Kullu	Seeds and gum	Gum is used as laxative and in dental fixture. It contains strcurensis & polysaccharides.	S

contd...

Table 11.6 - *contd...*

Scientific Name	Local Name	Part's Used	Uses and active chemical constituents	Status
Syzygium cumini	Jamun	Fruits and leaves	Usedful in diabetes, diarrhoea and dysentery. It contains quercetin and jamboline	PS
Tagetes erecta	Genda	Leaves and flowers	Leaf juice is used in earache. It yields an essential oil rich intagatone and other terpenes.	C
Tagetes patula	Genda	Leaves and flowers	Leaf juice is used in earache. It yields an essential oilrich in pipertenone, tagatone etc.	C
Tephrosia purpurea	Sarponkha	Whole plant	The plant is useful in constipation, worms, diarrhoea. it contains rutin and purpurin.	S
Terminalia arjuna	Arjuna	Bark	The bark is taken in cardiac disorders. Bark contains arjuline, arjunosides and tannin.	PS
Terminalia bellerica	Baahera	Fruits	It it one important ingredient of tirphala and useful in piles and fever. It contains beta-sitosterol, gallic acid & menitol.	PS
Tetminalia chebula	Harra	Fruits	The fruits powder is taken in constipation. The roasted fruits is useful in cough. It contains chebulin, terpenes and saponins.	PS
Thuja occidentale	Thuja	Leaves and cones	Useful in cough, fever, rheumatism and gout. It is astringent, digestive, laxative, diuretic and carminative. It yields thujone rich essential oil.	T
Thuja orientalis	Morpanki	Leaves and fruits	It is useful is hreumatism, amenorrhoea and carcinoma. It yields an essential oil rich in terpenes and 3 carene.	T

contd...

Table 11.6 - *contd...*

Scientific Name	Local Name	Part's Used	Uses and active chemical constituents	Status
Tinospora cordifolia	Gurbel	Stem and leaves	It is used as tonic, antiperiodic and aphrodisiac. It contains tinosporin, tinsoporol and tinosporic acid.	T
Tribulus terrestris	Chota Gokhru	Fruit	Fruits are aphrodisiac, tonic, diuretic and inflammatory. It contains diosgenin and steroidal saponin.	T
Tridax procumbens	Akdandi	Leaves	The juice of leaves applied in cuts and wounds including ulcers.	S
Tylophora indica	Anantmool	Leaves and roots	Leaves are taken in asthma and bronchial diseases. It contains tylophorone and tylophorinine.	T
Urginea indica	Jangali pyag	Bulb	It is used in heart trouble, cough, bronchitis. It contains scillarenes and glucosides.	S
Ventilago caliculta	Keoti	Seeds	Oil is used in rheumatism. Oil is eaten by tribal population. It contains fatty acids.	S
Vetiveria zizanioides	Khus	Roots	It is stimulant, diaphoretic and refrigerant. Useful in sprain and rheumatism. It contains essential oil rich in vetivone.	PS
Vitex negundo	Nirgundi	Leaves	The leaves are used in body pain and rheumatism. It contains artematin and casticin.	PS
Withania somnifera	Ashwa-gandha	Roots and leaves	Roots are taken to improve vigour and stamina. Useful in general weakness and rheumatism. It contains withanolides and withaferin.	PS

Table 11.7 : List of Medicinal plant becoming extinct and vulnerable in this region

Sl.No	Botanical Name	Vernacular Name	Parts used in various diseases
1	*Acorus calamus*	Bach	Rhizome are used as tonic and stammering.
2	*Argyreia spaciosa*	Samudra Sokh	Leaves are applied for boils and tumour.
3	*Aristolochia indica*	Ishwarmool	Roots are used for sanke bite.
4	*Boerhavi diffusa*	Punannava Lal	Roots cure corneal ulcer and cough.
5	*Celastrus paniculatus*	Malkangni	Seeds are used to improve memory and skin diseases.
6	*Chlorophytum borivillianum*	Safed Musli	Roots are used as tonic.
7	*Citrullus colocynthus*	Indryan Badi	Leaves smoke for asthma and for blackening hair.
8	*Curcuma angustifolia*	Tikhur	Rhizomes are good source of starch, nutritive
9	*Curcuma caesia*	Kali Haldi	Rhizomes are used for sprains and brusies.
10	*Dioscorea deltoidea*	Katalu	Tuber is used as tonic.
11	*Eclipta alba*	Bhring Raj	Roots are used with ajwain for enlargement of liver.
12	*Embelia ribes*	Baividang	Seeds are used to kill pare worms and round worms.
13	*Gloriosa superba*	Kalihari	Roots are used for white patches (Leucoderma)
14	*Glymnema sylvestris*	Gurmar	Leaves are used for diabetes with other drugs.
15	*Hedychium spicatum*	Kapoor kachri	Root is given for stomach ache.
16	*Litsea sebiferapers*	Maidalakdi	Bark for joining broken bones.
17	*Plumbago zeylanica*	Chitrak	Roots are used for rheumatism.
18	*Rauvolfia tetraphylla*	Chota Chand	The root is given for epilepsy.
19	*Rauvolfia serpentina*	Sarpagandha	Roots of the plant are given in hypertension.
20	*Smilax zeylanica*	Ram Datoon	Roots is taken for spermatorrhoea, weaknees.
21	*Tephrosia purpurea*	Sarpunkha	Roots are used as laxative and to treat worms.
22	*Tylophora indica*	Antamool	Leaves are used for asthma.
23	*Urginea indica*	Jangli Pyaj	Rhizomes are used in bronchitis.

Conservation Strategies

Due to manifold human interferences, the valuable medicinal and aromatic plants are becoming extinct. Commercial exploitation has resulted in the eradication of several important MAPs from their natural habital. The state of Madhya Pradesh is one of the major mega diversity centers in the country, which gives us greater responsibility to make efforts towards conservation of our richest biodiversity for future generations. In recent years, the conservation of medicinal plant has gained proper attention as their medicinal values are becoming more and more popular. At the initiative of the Chief Minister, the state Government has constituted the 'Madhya Pradesh State Bio-diversity Board' to protect its rich biological wealth and make it a substantial source of income for the rural and tribal communities. Some of the useful suggestions for the conservation of our biodiversity may be as follows:

1. Development of area specific agro-techniques for cultivation of MAPs as crop to take off pressure from natural wild stock.
2. Documentation of the location and ecological status of bioresources.
3. Development of awareness programme for sustainable use and conservation of valuable MAPs among rural people.
4. Domestication of MAPs.
5. Documentation of indigenous knowledge of utilization of medicinal plants.
6. Conservation in herbal/botanical gardens (ex-situ conservation).
7. In-situ conservation of MAPs by protecting their natural habitats by people participation.
8. Periodical workshop, training programmes for farmers and entrepreneurs to appraise recent development and advantage of cultivation of MAPs.

Satpura Plateau of Madhya Pradesh with diverse agro-climatic conditions, large biodiversity and strategic geographical location is likely to emerge as a leading producer and supplier of medicinal plants. The paper gives a resume of activities undertaken in recent years by Centre for Forestry Research and Human Resource Development, Chhindwara as a part of endeavour towards conserving the medicinal plant wealth and promoting the cultivation of medicinal plant. Training programmes on conservation and cultivation of medicinal plants have also been carried out as a part of Human Resource Development activity. Agrotechniques have been developed for the cultivation of Abelmoschus moschatus, Acorus calamus, Andrographis paniculata, Rauvolfia serpentina, Withania somnifera, Cymbopogon flexuosus, Cymbopogon moartinii and Mentha arvensis.

Source: A.K. Pandey *et al.,* in the Indian Forester, Vol 131 No. 7, July 2005–Modified from.

Chapter Twelve
Potential Drug Plants of Laterite and Arid Zones

In two separate lists important medicinal plant species have been listed. The list of Arid zone in particular may be considered as of ethnological importance.

The lists show that most of the plants are sporadic which shows absence of regeneration and lack of gregariousness of species. The species represent the large arid zone or drought prone zone of India.

Ethno Medicinal Plants in the Indian Arid Zone

Whatever be the assessment of the authors of the list it is a fact that most of the species occur sporadically and the inflated list must not give an impression that the area is rich in medicinal plants. Very few are presently safe. Only a few sporadic species occur in gregarious patches. Of the few species listed by Kumar *et. al.*, Aloe, Asparagus, Glycyrrhiza, Nardostachys are very much depleted, besides Vanda, Strichnos, Nelumbium, Gloriosa, Gymnema, Dioscorea, Costus, Bryonia are threatened in the area.

Droughts, over-exploitation and grazing are derogatory factors to their survival.

In the arid zone of Jaisalmer, Barmer, Bikaner and Jodhpur 682 species have been reported of which 131 sp. have been found to be of ethno-medicinal value. Of these, 41 species are collected and sold in the arid zone.

Some of the species are (There are other 60 sp.):

Acacia nilotica, Justicia adhatoda, Aloe vera, Asparagus racemosus, Azadirachta indica, Butea monosperma, Capparis decidua, Cassia angustifolia, Cassia fistula, Cassia tora, Citrallus lanatus, Clerodendrum indicum, Commiphora wightii, Cuscuta reflexa, Emblica officinalis, Evolvulus alsinoides, Glycyrrhiza glabra, Indigofera cordifolia, Ipomoea digitata, Jatropa curcus, Lawsonia enermis, Mucuna pruriens, Moringa oleifera. Nardostachya jatamansi.

(*Source* : Modified from S. Kumar *et al.*, Ind. For. Jany. 2005)

Table 12.1 : Ethno-medical Plants of Arid Zone

Scientific Name	Local Name	Status
Anthemis pyrethrum	Akor kora	S*
Argomone maxicana	Atkuti	C
Amorphophathus campanulatus	Atopinda	PS
Acacia nilotica	Babla	S
Alangium lamarckii	Dhela	S
Adiantum lunulatum	Dodhar	S
Asclepias rosea	Dudhiani	S
Acacia farnesiana	Gabur	PS
Amaranthus gangeticus	Gandhari	PS
Alternanthera sessiles	Garundi	T
Aristolochina indica	God	C
Areaca catechu	Gua	C
Anogeissus latifolia	Hesel	PS
Amaranthus spinosus	Janum arak	PS
Acacia intsia	Kondro	S
Agave americana	Konga	C
Aurea lanata	Lopon arak	S
Anona squamosa	Mandargon	C
Antidesma diandrum	Matha arak	T
Agle marmetos	Sinjo	C
Andropogon muricatus	Siram	S*
Alstonia scholaris	Chatnichal	S
Abrus precatorius	Kawet	S
Adina cordifolia	Karam	S
Artocarpus integrifolia	Kanthar	C
Asparagus racemosus	Kadar nari	T
Anthocephalus chinense	Kadam	PS
Borassus flabelliformis	Andiatale	S
Bursera serrata	Amru	S
Butea monosperma	Daremurap	PS
Bombax ceiba	Edel	PS
Barringtonia acutiangula	Hinjor	PS
Bassia latifolia	Koera	PS

contd...

Table 12.1 – *contd...*

Scientific Name	Local Name	Status
Brassica campestris	Lutnifuri	C
Bryonia lacinosa	Pond Kahubloke	T
Basella alba	Purgi	C
Blumea wightii	Bondoc	S
Bowsellia serrata	Salga	PS
Bridelia stipularis	Saudisaba	S
Buchanania latifolia	Tarop	PS
Boerhaavia repens	Ohok arak	S*
Buttneria herbacea	Ramraj	S
Bauhinia retusa	Jhinjit	S
Bonnaya veronicaefolia	Kadar	S
	Akaona	
Calotropis procera		S
C. gigantean		S
Cuscuta chinensis	Alagiri	S
Capparis horrida	Asaria	S
Combretum decandrum	Aten	S
Cephalandria indica	Atokundri	S
Clerodendrum siphonanthus	Barn	S
Calamus rotang	Bet	T
Cassia tora	Bheda deren	S*
Caesulia axillaries	Bhelaonja	S
Coriandrum sativum	Dhania	C
Cynodon dactylon	Dhubighas	PS
Crataeva nurvula	Ekasira	S
Cannabis sativa	Ganja	C
Coix lachryma jobi	Gargodi	S
Croton oblongifolius	Gote	S*
Cissus quadrangularis	Hadjora	C
Cleome viscosa	Hurhura	S
Citrus medica	Jambir	C
Caram ajowan	Jawar	C
Cucurbita moschela	Kehnda	C
Celastrus paniculatas	Kujri	S
Careya arborea	Kumbir	S
Commelina bengalensis	Orak siranre	S
Capsicum frutescens	Maric	C
Cajanus indicus	Raher	C
Casearia tomentosa	Ride	S
Clerodendrum serratum	Soramluter	A

contd...

Table 12.1 – *contd...*

Scientific Name	Local Name	Status
Colocasia antiquorum	Saru	C
Crinum Zeylanicum	Sikiom	S
Cyperus tegelum	Sura ghas	S
Cisampelos pareira	Tejomala	S
Canavalia ensiformis	Tihon	S
Curcuma angustifolia	Paroda	S
Costus speciosa	Orop	T
Cordia myxa	Bucchal	C
Cotylendon laciniata	Hemsagor	C
Casearia tomentosa		C
Cryptolepis buchanani	Ufri	C
Dillenia scabrella	Agaire	S
Dragia volubile	Andia kongal	S
Dioscorea crispate	Bayan	T
Datura alba	Datra	S
Dioscorea pentaphylla	Durason	T
Diospyrus montana	Gada Terel	S
Desmodium triflorum	Hudin catomaral	S
Dioscorea daemonum	Kolo	T
Dillenia indica	Kor Kot	PS
Diospyros embryopteris	Makar kenda	S
Dolichos lablab	Malhan	C
Dioscorea oppositifolia	Piska	T
Dolichos biflorus	Pond horee	S
Dillenia pentagyna	Sahar	S
Dalbergia latifolia	Satsayar	C
Delphinus gangetica	Suhako	S
Desmodium gangeticum	Toyobal	S
Dioscorea damonum	Koloda	T
Diospyrus montana	Gada terel	S
Emblica officinalis	Arakmeral	C
Embelia robusta	Bhabri	S
Entada scandens	Bidhanta	S
Euphorbia antiquorum	Etkec	S
Elleteria cardamom	Elachi	C
Enhydra fluctuans	Hemca aral	S
Eleusine coracana	Kodo	C
Eriosma chinensis	Konden	S
Emilia sonchifolia	Kutai lutur	S

contd...

Table 12.1 – *contd...*

Scientific Name	Local Name	Status
Erythrina indica	Marar baha	S
Erhetia laevis	Pusipan	S
Euphorbia hirta	Pusitoa	S
Euphorbia thymifolia	Nanha pusitoa	S
Elaeodendron roxburghii	Niuri	S
Ficus bangalensis	Bare	PS
Fimbristylis monostachya	Bindinutha	S*
Flemingia conjesta	Binbui	S
Ficus religiosa	Hesak	PS
Ferrula asafoetida	Hin	C
Feronia elephantum	Katbel	S
Ficus glomerata	Loa	S
Flacourlia ramontchi	Marlec	S
Gymnema hirsutus	Andia moron	D
Grewia sapida	Barka paker	S
Grislea tomentosa	Dhai	S
Gardenia turgida	Dhundukif	S
Gmelina arborea	Goda kasmar	C
Gardenia latifolia	Popro	C
Gloriosa superba	Selepsomanom	D
Grewia polygama	Setaandga	S
Grewia villosa	Tarsekolap	S
Gynandropsis pentaphylla	Seta kata	S
Gossypium arboreum	Kaskom	S
Hippocratea arborea	Bandlotanari	S
Hymenodictyon excelsum	BhorKond	S
Holarrhena pubescens	Birhal	PS
Hygrophila spinosa	Gukia	S*
Heliotropium indicum	Hate sunda	S
Helicteres isora	Petcambra	S
Heteropogon contortus	Sauri	PS
Hygrophila spinosa	Sirauna	S
Hibiscus canabinus	Soupal	S
Helianthus annus	Surujmukhi	C
Ichnocarpus frutescens	Dudhilota	S
Indigofera pulchella	Hutar	S
Ipomoea quamoclit	Kidinibaha	S
Ipomoea batatas	Sakarkenda	C
Ixora parviflora	Merom	C
Justicia adhatoda	Basok	S

contd...

Table 12.1 – *contd...*

Scientific Name	Local Name	Status
Jatropha curcas	Bhernla	S*
Jasminum arborescens	Hundbaha	
Limnophila roxburgiana	Akarbaha	S
Leucas chephaloides	Andiadhu	S
Luffa acutangula	Atojhinga	C
Lygodium flexuosum	Badgocak	C
Loranthus vestitus	Baru banda	S
Leea macrophylla	Dhalka hetkan	S
Leucas cephalotes	Dhurup arak	S
Lannea coromendelina	Doka	S
Leea hirta	Duria hatkan	S
Lipidium sativum	Halim	S
Leea aspera		S
Luffa acutangula	Porol jinglia	C
Lagerstroemia parviflora	Sekre	PS
Linum usitatissimum	Tisi	C
Lowsonia alba	Mendisakam	C/PS
Mezoneurum cuculatum	Baghin	S
Martynia diandra	Baglauca	S
Mimosops elengi	Bar	C
Mucuna pruriens	Etka	S
Millettia auriculata	Hehel	S
Marshalia quadrifolia	Maran catom	S
Mentha sativa	Pudin arak	C
Mallotus philippinensis	Rora	S
Mimosa rubicaulis	Sega	S
Monochoria plantaginea	Setapan	S
Morinda tinctoria	Tampurcaili	S
Mangifera indica	U1	C
Maringa oleosa	Munga chal	C
Michelia champaca	Champ	C
Musa paradisiaca	Kaera	
Nelumbium speciosum	Poraeni	D
Nerium odorum	Rajbaha	C
Nyctanthes arbor-tristis	Saparom	C
Nigella indica	Kalia jira	C
Nardostachys jatamansi	Nagranter	D
Oroxylum indicum	Banahalak	S
Ocimum basilicum	Bharbhari	C

contd...

Table 12.1 – *contd...*

Scientific Name	Local Name	Status
Ochna squarrosa	Birchampa	S
Ocimum basilicum var. thysiflorum	Dimbu Baha	C
Ougenia dalbergioides	Rot	S
Opuntia dillenii	Sapin	S
Ocimum sanctum	Tursi	C
Oldenlandia biflora	Khat pipra	S
Panicum foetida	Adagathia	S
Psidium guava	Amsopori	C
Pollinia eriopoda	Backom	S
Pentapetes phoenicia	Barebaha	S
Phyllanthus lanceolarius	Baria kandhum	S
Polygonum plebejum	Bhaya bhagua	S
Polygala chinensis	Gaighurarak	S
Plumeria acutifolia	Gulanj baha	C
Panicum miliare	Gundli	S
Physalis minima	Handikundi	S
Panicum crusgalli	Iri	S
Plantago ispaghala	Isabgul	C
Paspalum scrobiculatum	Janhe	S*
Polycarpaea corymbosa	Janhenanjom	S
Panicum flavidium	Jerajanhe	S*
Phoenix sylvestris	Khajur	C/PS
Pandanus odoratissimus	Kiabaha	S
Phoenix acualis	Kita	S
Plumbago zeylanica	Kitauri	S
Pongamia pinnata	Koranj	C/PS
Polygala crot alarioides	Lilkathi	S
Piper longum	Pipul jo	C
Phaseolus mungo	Ramra	C
Pueraria tuberosa	Tirra	S
Polygala chinense	Gaighura	S
Paspalum sp.		S
Pterospermum acerifolium	Mackunda	C
Randia uliginosa	Darependa	S
Racinus communis	Eradom	C
Ruellia suffruicosa	Caulia	S
Randia dumetorum	Loto	S
Smilax ovalifolia	Atkir	S
Sorghum vulgare	Bajra	C
Spatholobus roxburghii	Bandonari	S

contd...

Table 12.1 – *contd...*

Scientific Name	Local Name	Status
s.l.e.chera oleosa	Baru	S
Solanum melongina	Bengar	C
Solanum stramonifolium	Bengar betahel	C
Scindapsus officinalis	Dare japak	C
Sapindus detergens	Darka rista	S
Shorea robusta	Datauni	S
Solanum xanthocarpum	Edheranginijanum	S
Stephegyne parvifolia	Gore	S
Syzygium cumini	Kod	S
Strichnos potatorum	Kuela	D
Symplocos racemosa	Lodam	D
Syzyzium caryophyllata	Lorphul	C
Spermacoce hispida	Pitua arak	S
Scirpus rotundus	Putki	S
Smilax ovalifolia	Raepan	S
Soymida febrifuga	Ruhen	S
Streblus asper	Sahra	S
Sterculia colorata	Sesebaha	S
Semecarpus anacardium	Soso	S
Scirpus monstachyus	Sukri mutha ghas	S
Salix tetrasperma	Surukue	S
Sterculia urens	Telhec	S
Sygyzium operculatum	Totnopal	S
Spondius mangifera	Ambra	C
Solanum jacquini	Rangim janum	S
Sida humilis	Joka sakam	S
Terminalia tomentosa	Atnak	S/PS
Terminalia chebula	Hortoki	S
Themeda gigantea	Kus	S
Terminalia bellerica	Lopon	S
Tetranthera monopetala	Pojo	S
Tragia involuerata	Sondhaeni	S
Trigonella foenum	Milhi	S
Thespesia lampas	Bonkapasi	S
Triumfetta rhomboidea	Bhidijenetep	S
Terminalia arjuna	Kauha chal	C/PS
Urena sinuata	Bhidijanatep	S
Urtica pentandra	Khetpipra	S
Vernonia cinera	Bangaura	S

contd...

Table 12.1 – *contd...*

Scientific Name	Local Name	Status
Vitex alata	Bar	S
Vanda roxburghii	Dare banki	D
Viscum attenuatum	Dare katom janga	S
Vitis latifolia	Icewer	S
Zingiber zerumbet	Mahabari	C
Vernonia anthelmintica	Saukha	S
Vigna vexillata	Soroan	S
Vitex peduncularis	Sim kata arak	S
Vitex negundo	Sinduari	S
Ventilego calyculata	Bonga sarjom	S
Vitis tomentosa	Ghora ladaure	S
Wrightia tomentosa	Atkura	S
Woodfordia fruticosa	Gadaical	S
Wendlandia tinctorium	Tilgai	S
Zingiber officinalis	Adhe	C
Zehneria umbellata	Atda	S
Zizyphus jujuba	Janum	S
Zizyphus oenoplia	Kuriframa	S
Zehneria umbellata	Kundri	S
Zizyphus rugosus	Sekra	S
Zingiber cassumunar	Orsoren	S
Zizyphus xylopyra	Kaera	S

NB, Legends : S = Sporadic; S* = gregarious patches; C = cultivated; PS = Presently sage; T = Threatened

Table 12.2 : Laterite areas of West Bengal, Bihar and Jharkhand

Scientific Name	Status	Scientific Name	Status
Abelmoschus moschatus	*S*	*Achyranthes pavonina*	*S*
Abrus precatorius	*S*	*Adiantum lunulatum*	-
Abrus pulchellus	*S*	*Aegle marmelos*	*C/PS*
Abutilon indicum	*S*	*Aerva javanica*	*S*
Abutilon theopharasti	*S*	*Aerva lanata*	*S*
Acacia farnesiana	*S*	*Agave americana*	*C*
Acacia pennata	*S*	*Agave cantala*	*C*
Acacia rugata	*S**	*Agave sisalana*	*C*
Acalypha indica	*S**	*Ageratum conyzoides*	*S**

contd...

Table 12.2 – *contd...*

Scientific Name	Status	Scientific Name	Status
Ahlaia odoratissima	S	*Bixa orellana*	C
Alangium salvifolium	S	*Blumea lacera*	S*
Albizzia amara	S	*Boerhavia diffusa*	S*
Alocasia indica	C	*Breynia vitis*	S
Aloe barbadensis	C	*Bryonopsis laciniosa*	S
Alstonia scholaris	PS	*Bryophyllum pinnatum*	S
Alternanthera sessilis	S*	*Buchanania lanzan*	PS
Amaranthus viridis	S*	*Buettneria herbacea*	S
Ambroma augusta	S	*Butea superba*	S
Amorphophallus campanulatus	C	*Caesalpinia bonduc*	S
Amorphophallus sylvaticus	C	*Calotropis gigantean*	S
Ampelocissus latifolia	S	*Capparis zeylanica*	S
Andrographic paniculata	T	*Cardiospermum halicacabum*	S
Anisochilus camposus	S	*Carissa carandas*	C
Anisomeles indica	S	*Carissa spinarum*	C
Annona reticulata	C	*Cassia absus*	S
Annona squamosa	C	*Cassia alata*	S*
Antidesma acidum	T	*Cassia angustifolia*	C
Argemone mexicana	S	*Cassia fistula*	PS
Aristolochia India	T	*Cassia mimosoides*	S
Artabotrys odoratissimus	C	*Cassia occidentalis*	S*
Artemisia vulgaris	S	*Cassia tora*	S*
Asparagus adscendens	T	*Casuarina equisetifolia*	C
Asparagus officinalis	T	*Catharanthus roseus*	C
Asparagus racemosus	T	*Cayratia trifolia*	S
Atalantia missionis	S	*Ceiba pentandra*	C
Atylosia scarabaeoides	S	*Celastrus paniculatus*	S
Averrhoa carambula	C	*Celosia agrentea*	S
Azadirachta indica	C/PS	*Centella asiatica*	S
Azanza lampas	S	*Centratherum anthelminticum*	S
Bacopa monnieri	S	*Cereus pterogonus*	S
Barleria cristata	S	*Ceriscoides turgida*	S
Barleria lupulina	S	*Cheilanthes tenuifolia*	T
Barleria prionitis		*Cinnamomum camphora*	C
Barleria strigosa	S	*Cinnamomum tamala*	C
Bauhinia acuminata	PS	*Cinnamomum Zeylanicum*	C
Bauhinia racemosa	PS	*Cissus adnata*	S
Bauhinia vahlii	PS	*Cissus quadrangularis*	S
Belamcanda chinensis	S	*Citrullus colocynthis*	C

contd...

Table 12.2 – *contd...*

Scientific Name	Status	Scientific Name	Status
Cleistanthus collinus	*S*	*Desmodium gangeticum*	*S*
Cleome icosandra	*S*	*Desmodium motorium*	*S*
Clerodendrum indicum	*S*	*Desmostachya bipinnata*	*PS*
Clerodendrum phlomoides	*S*	*Dicliptera bupleuroides*	*S*
Clerodendrum serratum	*S*	*Dillenia indica*	*S*
Clerodendrum viscosum	*S**	*Dillenia pentagyna*	*PS*
Clitoria ternatea	*C*	*Dioscorea alata*	*T*
Coccinia grandis	*C*	*Dioscorea bulbifera*	*T*
Coccinia indica	*C*	*Dioscorea esculenta* Var. *spinosa*	*T*
Cochlospermum religiosum	*PS*	*Dioscorea pentaphylla*	*T*
Coleus amboinicus	*C*	*Dioscorea triphylla*	*T*
Colocasia esculenta	*C*	*Diospyros melanoxylon*	*S*
Colocasia nymphaeifolia	*C*	*Ecbolium viride*	*C*
Commelina benghalensis	*S*	*Eclipta prostrata*	*S*
Commiphora mukul	*T*	*Ehretia laevis*	*S*
Costus speciosus	*T*	*Elaeocarpus ganitrus*	*C*
Cotula anthemoides	*S*	*Elephantopus scaber*	*S*
Crataeva nurvala	*S*	*Elettaria cardamomum*	*C*
Crinum asiaticum	*C*	*Emblica officinalis*	*C/PS*
Crotalaria burhia	*C*	*Enydra fluctuans*	*PS*
Crotalaria pallida	*C*	*Erythrina variegata var. Orientalis*	*PS*
Crotalaria prostrata	*S*	*Eulophia nitida*	*S*
Crotalaria retusa	*S*	*Eupatorium triplinerve*	*S*
Cryptolepis buchanani	*S*	*Euphorbia antiquorum*	*S*
Curculigo orchioides	*T*	*Euphorbia hirta*	*S*
Curcuma amada	*C*	*Euphorbia nerifolia*	*S*
Curcuma aromatica	*C*	*Euphorbia thymifolia*	*S*
Curcuma caesia	*C*	*Euphorbia thymifolia*	*S*
Curcuma longa	*C*	*Euphorbia tirucalli*	*S*
Curuma zedoaria	*T*	*Euphoria longam*	*S*
Cymbopogon citrates	*C*	*Ficus glomerata*	*S/PS*
Cymbopogon martini	*C*	*Flacourtia indica*	*S/PS*
Cyperus kyllinga	*PS*	*Flacourtia jangomas*	*S/PS*
Cyperus rotundus	*PS*	*Flemingia chappar*	*S*
Cyperus triceps	*PS*	*Flemingia strobilifera*	*S*
Dalbergia sissoo	*PS*	*Garcinia xanthocymus*	*S*
Dalbergia volubilis	*S*	*Gardenia gummifera*	*S*
Datura innoxia	*S*	*Gardenia latifolia*	*S*
Datura metel	*S*		
Datura stramonium	*S*		

contd...

Table 12.2 – *contd...*

Scientific Name	Status	Scientific Name	Status
Gardenia resinifera	S	*Leucas plukenetii*	S
Gendarussa vulgaris	S	*Limnophila indica*	S
Gloriosa superba	T	*Limonia acidissima*	S
Gossypium herbaceum	S	*Lippia javanica*	S
Grangea maderaspatana	S	*Litsea glutinosa*	S
Grewia helicterifolia	S	*Ludwigia octovalvis*	PS
Grewia subinaequalis	S	*Lygodium pinnatifidum*	S
Gymenema sylvestre	T	*Mangifera indica*	PS
Gynandropsis gynandra	S	*Maranta arundinacea*	S
Habenaria commelinifolia	T	*Marsilea minuta*	S
Hedyotis corymbosa	S	*Martynia annua*	S
Helicteres isora	S	*Melastoma malabathricum*	S
Heliotropium indicum	S	*Melilotus alba*	S
Hemidesmus indicus	S	*Mentha piperita*	C
Hibiscus vitifolius	S	*Mentha spicata*	C
Holarrhena pubescens	PS	*Mesua ferrea*	S
Hybanthus enneaspermus	S	*Meyna laxiflora*	S
Hygrophila schulli	S	*Michelia champaca*	S
Hyptis suaveolens	PS	*Mimosa pudica*	S
Ichnocarpus frutescens	S	*Mimosa rubicaulis*	S
Indigofera tinctoria	S	*Mimusops elengi*	C/S
Ipomea sp	S	*Mimusops hexandra*	S
Ipomoea aquatica	PS	*Mollugo spergula*	S
Ipomoea paniculata	S	*Momordica dioica*	C
Ipomoea pes-tigridis	S	*Morinda citrifolia*	PS
Ipomoea quamoclit	S	*Morus alba*	C
Jatropha gossy ifolia	PS	*Mucuna pruriens*	S
Justicia adhatoda	PS	*Murraya koenigii*	PS
Kaempferia galanga	S	*Murraya paniculata*	PS
Kalanchoe pinnata	S	*Myristica fragrans*	S
Kirganelia reticulata	S	*Nelumbo nucifera*	T
Lannea coromandelica	S	*Nerium odorum*	S
Lantana camara	PS	*Nyctanthes arbortristis*	S
Lawsonia inermis	S	*Nymphaea alba*	T
Leea asiatica	S	*Nymphaea nouchali*	T
Leea macrophylla	S	*Nymphaea stellata*	T
Leonotis nepetaefolia	S	*Ocimum basilicum*	C
Leonurus sibiricus	S	*Ocimum canum*	S
Leucas cephalotes	S	*Ocimum gratissimum*	S

contd...

Table 12.2 – *contd...*

Scientific Name	Status	Scientific Name	Status
Ocimum kilimandscharicum	C	*Rauvolfia serpentina*	C
Ocimum sanctum	C	*Rauvolfia tetraphylla*	C/D
Ocimum tenuiflorum	S	*Rhomboidea*	
Opuntia dillenii	S	*Ricinus communis*	S
Oroxylum indicum	S	*Rivea hypocrateriformis*	S
Oxalis comrniculata	S	*Rungia pectinata*	S
Paederia scandens	S	*Sansevieria cylindrical*	C
Pandanus amaryllifolius	S	*Sansevieria roxburghiana*	C
Pandanus fascicularis	S	*Santalum album*	C/T
Passiflora suberosa	S	*Sapindus laurifolius*	S
Pedalium murex	S	*Saraca asoca*	C
Pentapetes phoenicea	S	*Schleichera oleosa*	S
Pergularia daemia	S	*Scindapsus officinalis*	C
Phlogacanthus thyrsiformis	S	*Scoparia dulcis*	S
Phyllanthus fratemus	S	*Selaginella rupestris*	S
Phyllanthus acidus	C	*Semecarpus anacardium*	S
Phyllanthus urinaria	S	*Sesbania grandiflora*	C
Phyllanthus virgatus	S	*Sesbania sesban*	C
Physalis minima	S	*Sida acuta*	S
Pimenta dioica	S	*Sida cordifolia*	S
Piper longum	C	*Sida humillis*	S
Piper nigrum	C	*Sida rhombifolia Linn. Var.*	S
Piper retrofractum	S	*Sida rhombifolia*	S
Plesmonium margaritiferum	S	*Smilax ovalifolia*	S
Plumbago indica	S	*Solanum indicum*	S
Plumbago zeylanica	S	*Solanum nigrum*	S
Plumeria acuminata	C	*Solanum sisymbrifolium*	S
Polygala crotalarioides	S	*Solanum surattense*	S
Portulaca oleracea	S	*Solanum torvum*	S
Premna herbacea	S	*Solena amplexicaulis*	S
Premna latifolia	S	*Sonchus oleraceus*	S
Pterocarpus marsupium	C	*Sphaeranthus indicus*	S
Pterocarpus santalinus	C	*Stachytrpheta jamaicensis*	S
Pterospermum acerifolium	C	*Stephania japonica*	S
Pueraria tuberosa	S	*Stereospermum suaveolens*	S
Punica granatum	C	*Streblus asper*	S
Putranjiva roxburghii	C	*Strychnos nux-vomica*	T
Quisqualis indica	PS	*Sunedrella nodiflora*	S
Randia spinosa	C	*Symplocos sp*	S

contd...

Table 12.2 – *contd...*

Scientific Name	Status	Scientific Name	Status
Syzgium operculatum	S	*Uraria lagopodioides*	S
Syzygium aromaticum	S	*Uraria picta*	S
Tamarindus indica	C	*Urena lobata*	S
Tectona grandis	C	*Urena sinuta*	
Tephrosia purpurea	S	*Vallaris solanacea*	S
Teramnus labialis	S	*Ventilago denticulate*	S
Terminalia arjuna	C	*Vernonia cinerea*	S
Terminalia bellirica	C	*Vetiveria zizanioides*	S
Terminalia chebula	C	*Vigna trilobata*	S
Thevetia peruviana	S	*Vitex negundo*	PS
Tinospora cordifolia	T	*Wedelia chinensis*	S
Tinospora sinesis	S	*Withania somnifera*	C/D
Tragia involucrate	S	*Woodfordia furticosa*	S
Trianthema portulacastrum	S	*Xanthium strumarium*	S
Tribulus terrestris	T	*Yucca gloriosa*	C
Trichosanthes tricuspidata	S	*Zingiber cassumunar*	S
Tridax procumbens	S	*Zingiber officinale*	C
Triumfetta rhomboidea	S	*Zingiber zerumbet*	S
Tylophora indica	S	*Zizyphus nummularia*	S
Typhonium Trilobatum	S	*Zizyphus oenoplia*	S

(*Source:* Modified from the publication of Govt. of West Bengal, Deptt. of Forests, Research Wing and other sources)

The work records 375 species in South Bengal; also covers laterite areas of Orissa and Jharkhand.

Legends : P = Planted; C = Cultivated; W = Wild; S = Sporadic; PS = Presently sage; D = Depleted; T = Threatened

N.B. of 375 species listed 55 sp. are cultivated, 58 species are planted and the rest 262 species are wild.

Comment

A list of 375 species have been prepared of which approximately 262 species one sporadic. Only a few among the sporadic species occur in large patches. As such about 78% of the species face arduous biotic and abiotic pressure.

Chapter Thirteen

Potential Medicinal Plants of North-East India

The primiary vegetation over bulk areas have been disturbed and modified and in some places destroyed by seismic activities, frequent landslides and resultant erosion. The activity of Man has led to the irreversible transformation in the landscapes and has resulted in colossal loss of biodiversity. Human influences have pushed many species to the brink of extinction and have caused havoc to natural fragile ecosystem. There has been decrease of forest cover of about 1800 sq. km. between 1991 and 1999 (F.S.I. 2000). Quality of forests has deteriorated to scrub.

North East India–a bowl of plant diversity has been identified as Hot Spots and Mega-diversity area due to its unique ecological setting and the center of meeting point of temperate and tropical flora. palaeo-arctic flora of Tibetan high land and wet-evergreen and rain forests flora of South East Asia and Yunan.

A perusal of the book entitled "Biodiversity Characterization" at Landscape Level in North-East India using "Satellite Remote Sensing and Geographic Information System" by Indian Institute of Remote Sensing, Deptt. of Space, Govt. of India 2002, may reveal that-

- Most of the States have extensive area under Shifting cultivation with short cycle.
- Heavily eroded area all over.
- Deforestation going on our large areas.
- Increase of human population and demand for space and forest produce.
- Cultivation made on sleep slope.

- Going poorer in biodiversity.
- Heavy mining operation.

(*Source:* Wildlife Institute of India, 2003)

The aforesaid publication of Indian Institute of Remote Sensing has recorded some information on the medicinal plants of India.

Although all the States have their record of medicinal plants the figure of R.S. Institute being the latest has been summarized as follows:

Arunachal Pradesh

About 419 species have been recorded to have medicinal use, some of which are, *Abies webbiana, Aconitum ferox, Alstonia scholaris, Aristolochia platanifolia, Artemisia vulgaris, Coptis teeta, Pongamia pinnata, Terminalia bellirical.* The author has discussed the status of some of the plants in lster in this Chapter.

Assam

228 specis have been listed of which importance of Alstonia scholaris Zanthoxylum nitidum Andrographis paniculata, Oroxylum indicum, Clerodendrum indicum, Jatropha curcas, Costus specious, Melastoma melabathricum have been mentioned specifically.

Manipur

85 species have been mentioned of which special mention has been made of Clerodendrum serratum.

Meghalaya

74 species have been mentioned of which special mention has been made of Clerodendurm serratum.

Mizoram

83 species have been mentioned of which special mention has been made of Alstonia scholaris and Lycopodium calvatum.

Nagaland

86 specis have been mentioned of which special mention have been made of Alpinia galanga.

Tripura

73 species have been mentioned of which special mention has been made of Azadirachta indica and Moringa oleifera.

Sikkim

70 species have been mentioned of which special has been made of Gmelina arborea and Pieris ovalifolia.

Besides, the foresters have listed medicinal plants occurring in various divisions in the respective working plans. The ethnobiologists have also listed plants used by the ethnic tribals of various states. Obviously the list would be a elaborate one to discuss on these plants. So a few important species have been selected for the present reward.

The status is representsed by legend/abbreviations.

P.S.	=	Presently safe	C	=	Cultivated
D	=	Depleted	S	=	Sporadic

The legend sporadic (S) has to be properly evaluated. It means at the species occur here and there, sometimes in patches, but not abundantly. This legend also indicates that the species may be eatened at undesireable habits.

The legend 'Cultivated' C indicates that protection has given to the species considering the medicinal importance and rarely of the species.

In reality the herbs and shrubs in India find protection in inaccessible areas only and to some extent in the protected areas. At other places biotic factors play a derogatory role in their depletion. These depleted plants at last find shelter on the marginal lands, human habitation premises and sporadically here and there only to survive as they have been eradicated from the habitats where they once flourished.

Observation on potential drug plants of India, listed by Chopra, Kirtikar and Bose and others has been made by the present in their book entitled "Biodiversity Engangered" (2002).

Trees (Need protection– A few selected species)

Balsamodendron mukul

Bixa orillana

Camphora officinarum

Callophyllum inophyllym

Caryota urens

Ceiba pentandra

Juglans regia

Cupressus sempervirens

Betula utilis

Aquillaria malaccensis

Taxus baccata

Juniperus communis

Screbera switeniodes (Ghanta parul).

Medicianal Plant Trees Quantitative Studies

An extract from voluminous field survey (quantitative data *i.e.,* per ha.. sp. number) on a few areas is reproduced. From the data any body can form a reliable figure on the status of a few tree species having medicinal values. Similarly, type studies, on statistical basis, have to be done on herbs, shrubs, climbers etc. to determine the status of medicinal plants of India and N.E. India.

Sikkim

Figures of South-Western Sikkim are only available.

Prime species are- Symplocos theifolia (Frequency occurrency is 90) and stems (per ha. are 69) has the maximum density; this species is of immense medicinal value.

13.1 : Scores Recorded by Remote Sensing Institute

Botanical Name	Status	Botanical Name	Status
Abies webbiana	PS	Bacopa monniera/monnieria	T
Abroma augusta	S	Balanites aegyptiaca	T
Abrus precatorius	T	Baliospermum montamum	T
Abutilon indicum	S	Baliospermum oxillare	T
Acacia arabica	PS	Balsamo dendron mukul	T
Acacia catechu	PS	Bambusa arundinacea	C/S*
Acacia farnesiana	PS	Bambusa balcooa	C
Acacia leucophloea	PS	Bambusa bamos	C
Acacia suma	PS	Bambusa spinosa	
Achyranthes aspera	PS	Bambusa tulda	C
Aclypha indica	S	Bambusa vulgaris	C
Acorus calamus	T	Barleria cristata	S
Adenanthera pavonina	S	Barleria prionitis	S
Adhatoda vasica	PS	Barleria strigosa	S
Adiantum capilus	T	Barringtonia acutangula	S
Adiantum caudatm	T	Barringtonia racemosa	S
Adiantum lunulatum	T	Basella alba	C/PS
Adina cordifolia	PS	Basella rubra	C/PS
Aegle marmelos	PS	Bassia longifolia	C/PS
Aeschynomene aspera	S	Bauhinia purpuea	C/PS
Aeschynomene indica	S	Bauhinia racemosa	PS
Aganosma caryophyllata	S	Bauhinia variegata	PS
Aonitum ferox	T	Berberis aristata	T
Areca triandra	S*	Berberis asiatica	T
Argemone mexicana	S	Betula utilis	S
Argyreia speciosa	S	Bixa orellana	C
Aristolochia indica	T	Blepharis edulis	S
Artemisia maritima	S	Blumea lacera	PS
Artemisia vulgaris	S	Boerhaavia diffusa	PS
Artocarpus heterophyllus	C	Bombax eeiba	PS
Artocarpus lokoocha	S	Borasus flabellifer	PS
Arum trilobatum	S	Boswellia serrata	PS
Arundo donax	PS	Botanical Name	Status
Asparagus racemosus	T	Botanical Name	Status
Asterocantha longifolia	S	Brassica campestris	C
Astragalus candolleanus	S	Brassica juncea	C
Astragalus leucocephalus	S	Brassica napus	C
Averrhoa carambola	C	Bryonia laciniosa	T
Azadirachta indica	C	Buchanania lanzan	S

contd...

Table 13.1 – *contd...*

Botanical Name	Status	Botanical Name	Status
Butea monosperma	PS	Celosia cristata	C
Cadrella toona	PS	Centella asiatica	PS
Caesalpinia bonducella	S	Cephalandra indica	T
Caesalpinia pulcherrima	C	Chenopodium album	S
Cajanus indicus	C	Chenopodium ambrosioides	S
Calamus vininalis	S	Cicer arietinum	C
Callicarpa marcrophylla	PS	Cimicifuga foetida	S
Calophyllum inophyllum	S	Cinnamomum camphora	C
Calotropis gigantia	PS	Cinnamomum zeylanicum	C
Calotropis procera	PS	Cinnamonum tamala	C/S
Camellia drupifera	S	Cissus quadrangularis	S
Camellia japonica	S	Citrallus vulgaris	C
Camellia kissi	S	Citrus aurangifolia	C
Camellia sinensis	C	Citrus decumana	C
Camellia thea	C	Citrus limetoides	C
Camellia theifera	C	Citrus limon	C
Canabis sativa	PS	Citrus maxima	C
Canscora decussata	S	Citrus medica	C
Capparis decidua	S	Citrus reticulata	C
Capparis zeylanica	S	Citrus sinensis	C
Capsicum annum	C	Cleome pentaphylla	S
Capsicum frutescens	C	Clerodendrum indicum	S/PS
Cardispermum halicacabum	S	Clerodendrum serratum	PS
Careya arborea	S	Clerodendrum viscosum	PS
Carica papaya	C	Clitoria ternatea	C
Carissa caradas	C	Coccinia indica	S
Carthamus tinctorius	C	Cocculus hirsutus	PS
Carum carvi	C	Cochlospermum religiosum	PS
Carum copticum	C	Cocos nucifera	C
Cassi sophera	S	Coix aquatica	PS
Cassia alata	S	Coix gigantea	PS
Cassia fistula	S	Coix lachryma jobi	PS
Cassia occidentalis	S	Commelina benghalensis	S
Cassia tora	S	Commelina salicifolia	S
Catharanthus roseus	C	Convolvulus arvensis	
Cedrus deodara	C	Coptis teeta	T
Ceiba pentandra	C	Corchous capsularis	C
Celastrus paniculatus	S	Cordia dichotoma	S
Celosia argentea	C	Coriandrum sativum	C

contd...

Table 13.1 – *contd...*

Botanical Name	Status	Botanical Name	Status
Costus speciosus	T	Echinochloa frumentacea	PS
Crocus sativus	C	Eclipta alba	PS/S
Crotalaria juncea	C	Elaeocarpus ganitrus	T
Croton tiglium	S	Elephantopus scaber	S
Cucumis melo	C	Elettaria cardamomum	C
Cucumis sativa	C	Embelia ribes	S
Cucumis utilissimus	C	Emblica officinalis	C/PS
Cuminum cyminum	C	Enhydra fluctuans	PS
Curculigo orchioides	T	Ephedra gerardiana	T
Curcuma amada	S	Ephedra vulgaris	T
Curcuma aromatica	C	Eriodendron anfractuosum	C
Curcuma domestica	C	Ervatamia coronaria	C
Curcuma longa	C	Erythrina variegata	PS
Curcuma zedoraria	C	Eupatorium ayapana	S
Cuscuta reflexa	S	Eupatorium triplinerve	PS
Cymbopogon jwarancusa	S	Euphorbia hirta	PS
Cymbopogon schoenanathus	C	Euphorbia neriifolia	C/PS
Cynodon dactylon	PS	Euphorbia prostrata	
Cyperus rotundus	PS	Evolvulus alsinoides	PS
Dalbergia sissoo	PS	Evolvulus nummularius	PS
Damemia extensa	S	Feronia elephantum	PS
Datura metal	PS	Feronia limonia	PS
Daucus carata var. sativa	C	Ferula foetida	C
Delonix regia	PS	Ficus bengalensis	PS
Dendrocalamum hamiltonii	PS	Ficus carica	PS
Dendrocalamus falcata	PS	Ficus cunia	PS
Dendrocalamus strictus	PS	Ficus heterophylla	PS
Desmodium gangetcum	S	Ficus hispida	PS
Desmodium triflorum	S	Ficus infectoria	S
Desmostachya bipinnata	PS	Ficus racemosa	S
Dillenia indica	PS	Ficus religiosa	PS
Dioscorea bulbifera	T	Ficus rumkphii	S
Dioscorea jacquemontii	T	Flacourtia cataphracta	S
Dioscorea pentaphylla	T	Flacourtia indica	S
Dioscorea triphylla	T	Flacourtia jangomas	S
Diospyros peregrina	T	Fleurya interrupta	S
Dolichos biflorus	C	Foeniculum vulgare	C
Dolichos lablab	C	Fritillaria cirrhosa	T
Drynaria quercifolia	PS	Fritillaria roylei	T

contd...

Table 13.1 – *contd...*

Botanical Name	Status	Botanical Name	Status
Fumaria indica	S	Ipomoea batatus	C
Fumaria parviflora	S	Ipomoea quamoclit	C
Garcinia indica	S	Ipomoea reptans	C
Garcinia mangostana	S	Jasminum auriculatum	C
Garcinia morella	S	Jasminum gandiflorum	C
Garcinia pedunculata	S	Jasminum heyneana	C
Garcinia tintoria	S	Jasminum multiflora	C
Gardenia gummifera	S	Jasminum pubescens	C
Gardenia lucida	T	Jasminum sambac	C
Glycosmis pentaphylla	PS	Juglans regia	T
Gmelina arborea	PS	Juniperus communis	T
Gomphrena globosa	PS	Juniperus macropoda	T
Gossypium herbaceum	PS	Jussiaea repens	PS
Grangea maderaspatana	S	Justicia gendarussa	C/PS
Grewia asiatica	PS	Kalanchoe pinnata	C
Grewia tillaefolia	-	Laccifer lacca	C
Gymnema sylvestre	T	Lagenaria vulgaris	C
Gynandropsis gynandra	PS/S	Lannea grandis	PS
Gynocardia odorata	S	Laportea crenulata	PS
Habenaria edgeworthii	T	Lathyrus sativus	C
Habenaria latilabria	T	Lawsonia inermis	C
Heacteres isora	PS	Leucas cephalotes	S
Heliotropium indicum	S	Lilium polyphyllum	C
Hemidesmus indicus	D	Lilium tigrinum	C
Hemigraphis hirta	D	Limnanthemum cristatum	PS
Hibiscus abelmoschus	PS	Linum usitatissimum	C
Hibiscus esculentus	C	Lippia nodiflora	PS
Hibiscus mutabilis	C	Lobelia inflata	C
Hiptage benghalensis	C	Loranthus falcatus	S
Hiptage madablata	C	Loranthus longiflorus	S
Holarrhena antidysenterica	PS	Luffa acutangula	C
Hordeum vulgare	C	Luffa amara	C
Hydnocarpus kurzii	S	Luffa echinata	C
Hydnocarpus wightiana	S	Madhuca indica	C
Hygroyza aristata	PS	Mallotus philippinesis	PS
Hyoscyamus niger	T	Mangifera indica	C
Ichnocarpus freutescens	S	Marsilea minuta	PS
Imperata cylindrical	PS	Marsilea quadrifolia	PS
Indigofera tinctorea	S	Melia azedirachta	PS

contd...

Table 13.1 – *contd...*

Botanical Name	Status	Botanical Name	Status
Melilotus indica	S	Nyctanthes arbortristis	C
Melothria heterophylla	S	Nymphaea nouchali	T
Mentha spicata	C	Nymphaea rubra	T
Merremia tridentata	S	Nymphaea sp.	T
Mesua ferrea	PS	Nymphaea stellata	T
Michelia champaca	PS	Ochrocarpus longifolius	S
Mimosa pudica	S	Ocimum basilicum	C
Mimusops elengi	C	Ocimum kikimandscharicum	C
Mimusops hexandra		Ocimum sanctum	C
Mirabilis jalapa	C	Oldenlandia corymbosa	S
Mollugo hirta	S	Oroxylum indicum	S
Mollugo sperula	S	Oryza fatua	C
Momordica charantia	C	Oryza sativa	C
Momordica cochinchinensis	C	Ougeinia oojeinensis	S
Momordica muricata	C	Oxalis acetosella	S
Morinda citrifolia	S	Oxalis corniculata	S
Morinda tinctoria	S	Paederia foetida	T
Moringa oleigera	C	Pandanus odoratissimus	S
Morus alba	C	Pandanus tectorius	S
Morus atropurpurea	-	Panicum frumentacea	S
Morus indica	C	Paris polyphylla	
Morus laevigata	C	Paspalum scrobiculatum	PS
Mucuna prurita		Pergularia extensa	S
Murraya koenigii	C	Peripioca aphylla	-
Murraya paniculata	S	Phalogacanthus thyriflorus	S
Musa paradisiaca	C	Phaseolus radiatus var. attrea	C
Musa sapientum	C	Phaseolus radiatus var. grandis	C
Myrica nagi	T	Phaseolus radiatus	C
Myristica fragrans	T	Phaseolus sublobatus	C
Myristica malabarica	S	Phoenix aculis	PS
Nardostachys jatamansi	T	Phoenix dactylifera	PS
Nauclea cordifolia	S	Phoenix patudosa	PS
Nelumbo nucifera	T	Phoenix sylavestris	PS
Nephelium longana	S	Phragmites karka	PS
Nerium indicum	C	Phylantus simplex	S
Nerium odorum	C	Phyllanthus niruri	S
Nicotiana plumbaginifolia	S	Physalis minima	S
Nicotiana tabacum	C	Pintacia integerrima	T
Nigella sativa	C	Pinus longifolia	C

contd...

Table 13.1 – *contd...*

Botanical Name	Status	Botanical Name	Status
Piper aurantiacum	C	Purnus cerasoides	T
Piper betle	C	Putranjiva roxburghii	C
Piper cubeba	T	Pyrus communis	C
Piper longum	T	Quamoclit pinnata	C
Piper nigrum	T/C	Quisqualis densiflora	C
Pisum sativum	C	Quisqualis indica	C
Plumbago zeylanica	S	Randia dumetorum	S
Plumeria acutifolia	C	Ranuculus sceleratus	S
Plumeria rubra	C	Raphanus sativus	C
Podophyllum hexandrum	C/D	Rauwolfia canescens	
Podophyllym emodi	C/D	Rauwolfia serpentina	T
Poinciana pulcherrima	C	Rauwolfia tetraphylla	I
Poinciana regia	C	Rheum emodi	D
Polyalthia suberosa	S	Rhododendron arboreum	T
Polyanthes tuberosa	C	Rhus succedanes	S
Polyasthia longifolia	C	Richinus communis	C
Polygonatum cirrhifolium	C/D	Rosa damascena	C
Polygonatum oppositifolium	C/D	Roscoea purpurea	S
Polygonatum verticillatum	S	Rubia cordifolia	S
Polygonum hydropiper	S	Saccharum benaglense	PS
Polygonum orientate	S	Saccharum officinarum	C
Polypodium quercifolium	S	Saccolabium papillosum	S
Pongamia pinnata	S	Salvadora oleoides	
Portulaca oleracea	S	Salvadora persica	
Portulaca quadrifida	S	Sansevieria roxburghiana	C
Potamogeton indicus	PS	Santalum album	C
Premna integrifolia	S	Sapindus trifoliatua	S
Premna latifolia	S	Saraca indica	C
Prunus amygdalus	T	Sarcosttemma acidum	S
Prunus communis	C	Saussurea appa	T
Prunus persica	C	Schleichera oleosa	S
Prunus puddum	C	Schleichera trijuga	S
Psoralia corylifolia	C	Scindapsus officinalis	C
Psoralia corylifolia	S	Semecarpus anacardium	C
Pterocarpus marsupium	C	Sesbania grandiflora	C
Pterocarpus santalinus	C	Sesbania sesbams	C
Pterospermum suberifolium	S	Setaria italica	C
Pueraria tuberosa	S	Shcrebera swietteniolides	S
Punica grantum	C	Shorea robusta	

contd...

Table 13.1 – *contd...*

Botanical Name	Status	Botanical Name	Status
Sida cordifolia	S	Tamarix dioica	S
Sida rhomboidea	S	Tamarix gallica	S
Smilax glabra	T	Tamarix indica	S
Smilax indica		Tamarix troupii	S
Solanum indicum	S	Taxus baccata	T
Solanum khasianum	C	Tectona grandis	C
Solanum melongena	C	Tephrosia purpurea	S
Solanum nigrum	S	Terminalia arjuna	C
Solanum torvum	S	Terminalia belerica	PS
Solanum xanthocarpum	S	Terminalia chebula	S
Sorghum vulgare	C	Terminalia citrina	S
Soymida febrifuga		Thalictrum foiolosum	T
Sphaeranthus indicus	S	Thespesia populnea	S
Spinacia oleracea	C	Thevetia nerifolia	C
Spondias dulcis	S	Thevetia peruviana	C
Spondias mangifera	S	Tinospora cordifolia	T
Stehania glabra	S	Tinospora malabarica	T
Stephania hernandifolia	S	Tinospora tomenstosa	T
Stereospermum suaveolens	PS	Trachyspermum ammi	T
Stereospermum tetragonum	PS	Tragia involucrata	S
Streblus asper	S	Trapa bispinosa	PS
Strychnos nuxcomica	S	Trewia nudiflora	S
Strychnos potatorum		Trianthema monogyna	S
Swertia chirata	T	Trianthema portulacastrum	S
Sylvinia cucullata	S	Tribulus terrestris	T
Symplocos laurina	T	Trichosanthes anguina	C
Symplocos racemosa	T	Trichosanthes bracteata	C
Syzygium aromaticum		Trichosanthes dioica	C
Syzygium cumini	S	Trichosanthes palmate	C
Syzygium fruticosa	S	Trigonella corniculata	C
Syzygium operculatum	S	Trigonella foenum-graecum	C
Tabernaemontana coronaria	C	Triticum aestivum	C
Tacca aspera	S	Tylophora asthmatica	T
Tacca integrifolia	S	Tylophora indica	S
Tamarindus indica	C	Typha elephantina	PS
Tamarix aphylla	S	Typha latifolia	PS
Tamarix arthiculata	S	Typhonium trilobatum	S

Chapter Fourteen
Research and Cultivation

The chapter gives list of more than 50 Institutions engaged in Research and Cultivation of medicinal plants in India.

Not a single publication of the institution or cultivation authorities has made any assessment on the status of medicinal plants in India pertaining to density, frequency and regeneration status of various species.

It is suggested that these organizations set up research units for field survey of herb, shrub, climbers and grasses on statistical background. Research in this field is still very inadequate.

Table 14.1 : List of Some Institutions/Organisations Engaged in Research/Cultivation of Medicinal Plants

Sl. No.	Name of Institution	Agro-technique available
1.	Department of Agricultural Botany and Crop Physiology, Jawaharlal Nehru Krishi Vishwa Vidyalaya, JABALPUR-482 004 (MP)	–
2.	Regional Research Laboratory, BHUBANEWSWAR-751 013 (Orissa)	–
3.	Centre for Advanced Study in Botany, University of Madras, CHENNAI-600 025	Gymnema sylvestre (madhunashni) Andrographis Peniculata (Kalmegh)
4.	Kerala Agricultural University, Aromatic & Medicinal Plants Research Station, Asamannoor P.O. Odakkali, KERALA-683 549	Saraca asoca (Ashok)
5.	National Botanical Research Institute Rana Pratap Marg, LUCKNOW-226 001	

contd...

Table 14.1 – *contd...*

Sl. No.	Name of Institution	Agro-technique available
6.	Division of Floriculture, Medicinal & Agricultural Science and Technology, Shalimar, SRINAGAR- 191 121	Inula racemosa (Pushkarmool), Swertia chirata (Chirayata)
7.	Nagarjun Medicinal Plants Garden, Dr. Punjabrao Deshmuk Krishi Vidyapeeth, P.O. Krishinagar, AKOLA-4 (Maharashtra).	Embelia ribes (Vidanga)
8.	Tropical Botanical Garden and Research Institute (TBGRI), Karimancode, P.O. Palode, Thiruvanthapuram- 695 562 (KERALA)	
9.	Deptt. Horticulture & Project, Narender Dev University of Agriculture & Technology, Narander Nagar, P.O. Kumarganj, FAIZABAD-224 229	
10.	Central Institute of Medicinal and Aromatic Plants (CIMAP) P.O. CIMAP, LUCKNOW-226 015	
11.	Division of Plant Science & Ecology Regional Research Laboratory JORHAT-785 006 (Assam)	Chlorophytum arundinaceum (Musali Safaid)
12.	Head, Department of Agro-forest and Environment, H.P. Krishi Viswa Vidyalaya, PALAMPUR-176 062 (H.P.)	
13.	Department of Natural Products, Education & Research, Sector-67, S.A.S. Nagar, MOHALI-160 062 (Punjab)	Bacopa mannieri (Brahmi), Asparagus adscendensh
14.	Jamia Hamdard, Hamdard Nagar, NEW DELHI-110 062	Tinospora cordifolia (Guduchi)
15.	High Altitude Plant Physiology Research Centre, H.N. B. Garhwal University, Post Box -14, Srinagar, GARHWAL-246 174	Picrorrhiza kurroa, Aconitum heterophyllum (Atees) Nardostachys Jatamansi (Jatamansi)
16.	Herbal Garden, Herbarium & Research Institute in ISM, Manali-Pathankot Highway, Government of H.P., JOGINDER NAGAR District Mandi- 176 061 (HP)	Bacopa monnieri (Brahmi)
17.	NWFP, Division Tropical Forest Research Institute, P.O. RFRC, Mandla Road, JABALPUR-482 021 (MP)	

contd...

Table 14.1 – *contd...*

Sl. No.	Name of Institution	Agro-technique available
18.	Deptt. of Botany, J.N. Vyas University, JODHPUR- 342 001	Commiphora wightii (Guggal)
19.	Director, State Forest Research Institute, Polopather, JABALPUR- 482 008 (MP)	
20.	Horticulture (M& AP), University of Agricultural Sciences, G.K.V.K., Campus, BANGALORE-560 065	
21.	Head, NWFP, forest Research Institute (ICFRE), P.O. New Forest, DEHRADUN-248 006 (Uttaranchal)	
22.	Insttl of Himalayan Bioresource Technology, Palampur Post Box No. 6, HIMACHAL PRADESH-176 062	
23.	NBPGR, Pusa Campus, New Delhi-110 012	Asparagus racemosus (Satavari)
24.	NBPGR, Regional Station, Distt. Nainital BHOWALI-263 132 (UP)	
25.	NBPGR Regional Station, Phagli, SHIMLA- 171 004	Aconitum palmatum (Partivisha), Aconitum ferox (Vatsnab)
26.	Regional Research Laboratory (Jorhat) Branch, Ita Nagar, P.O. Naharlagun NAHARLAGUN-791 110 (Arunachal Pradesh)	
27.	Director, Indian Inst. Of Horticultural Research, BANGALORE-560 089	–
28.	Department of Agronomy, College of Agriculture, G.B. Pant University of Agriculture & Technology, PANT NAGAR-263 145 (UP).	
29.	Director, Regional research Laboratory (CSIR), Canal Road, JAMMU-TAWI-180 001 (J&K)	
30.	Mahatma Phule Krishi Vidyapeeth, Rahuri, Distt. Ahmednagar, MAHARASHTRA	
31.	NBPGR, Regional Station New Kench's Trace, Shillong, SHILLONG-793 013. (Meghalaya).	

contd...

Table 14.1 – *contd...*

Sl. No.	Name of Institution	Agro-technique available
32.	Deptt. of Horticulture, S.K.N. College of Agriculture, Rajasthan Agriculture University, JOBNER- 303 329 (Rajasthan).	
33.	Pt. Jawaharlal Nehru College of Agriculture & Research Institute KARAIKAL-609 603 (Pondicherry)	
34.	J.L. Nehru Ayurvedic Medicinal Plants Garden, Kothrud, PUNE (Maharashtra)	Asparagus racemosus (Shatavari)
35.	Uttan, Centre for Sustainable Development & Poverty Alleviation 18-A, Auckland Road, Civil Lane ALLAHABAD	Phyllanthus amarus (Bhumi amlaki), Asparagus racemosus (Shatavari), Bacopa Monnieri (Brahmi), Withania Somnifera (Ashwagandha)
36.	Survey of Medicinal Plants Unit Regional Research Institute of Unani Medicine, Post Box 70, Aligarh-202 001.	
37.	Guggal herbal form Mangliawas CCRAS, Ajmer (Rajasthan).	
38.	Department of Botany, J.N. Vyas University, Jodhpur- 342 001 Rajasthan.	
39.	Department of Plants Breeding, Vhaudhary Charan Singh Agriculture University Haryana, Hissar- 125 004.	
40.	CEDMAP, 60, Jail Road, Jahangirabad, Bhopal (Madhya Pradesh).	

(*Source* : National Medical Plants Board. Ministry of Health & Family Welfare, Government of India).

Research and Development Centres:

Following are R & D centers in India:

1. Council of Scientific and Industrial Research:
    - R.R.L. at Jorhat.
    - R. R. L. at Bhubaneswar.
2. Central Institute of medicine and Aromatic plants at Lucknow.

- I.C.A.R. Research:
- National Bureau of Plant Genetic Research.
- Indian Horticultural Research Institute.
- Cental Arid Zone Research Institute, Jodhpur.
- Central plantation Crop Research Institute, Kasaragod.

3. Universities:

- G.B. Plant University of Agriculture and Technology.
- Punjab Agricultural University, Ludhiana.
- Tamilnadu Agricultural University, Coimbatore.
- Kerala Agricultural University, Oddakali.
- Gujarat Agricultural University, Anand.

4. Others:

- Indian Council Of Forestry Research and Education, Dehradun.
- Cinchona Departmet

Medicinal plant directorates of West Bengal and Tamilnadu carry out research. to a limited edient or several servies.

The research should pinpoint the following points :

- Which part or parts are used: Is froot or root bark leaf, lower, fruit or any other part.
- Which time of the year such parts are to be collected to get best efficacy.

Cultivation

Details of cultivation have not been brought under the discussion. Only some important aspects have been discussed. Some aspects are-

- National Medical Board gives priority in cultivation of the following speices :

  Emblica officinalis, Saraca asoca, Withania somnifera, Aconitium heterofphyllum, Aegle marmelos, Phyllanthus amarus, Bacopa morneri, Santalum album, Swertia chirata, Tinospora cordifolia, Gymnema sylvestre, Commifera wightii, Plantago ovata, Nardostachys jatamansi, Glorosa superba, Andrographis paniculata, Garcinia indica, Saussurea costus,

Picorrhiza kurroa, Glycyrrhiza glapra, Solanum nigram, Chlorophytum boviisllianum, Cotus barbatus, Piper longifera, Belenis artolata, Crocus sativus, Rauwolfia superntina.

- Cultivation of Aromatic Plants:

  Pandanus fasciculatus, Aquilaria malaccansis, Cardamum, Clove, Eucalyptus, Fennel, Cinnamomum, Coriandum, Jasmine, Rose, Pepper, Vetiver, Nutmeg, Citronella, Belladona, Anoniluim, Acuminata, Gloriosa superba, Rauwolfia serpentia, Cephalis, Amnimajus, Pyrethrum, Chry santhemum, Cinerariae olium, Cinchona officinalis, C. micrantha, C. robusta, C. ledgerira, Aloe vera, Dioscorea floribunda, D. composita, Solanum khasianum, Costus speciosus, Datura innoxa, Datura metel, Datura stramonium, Solanum khsianum, Cassia absus, Cassia fistula, Celastrus zeylanious, Cissus quadrangularis, Coptis teeta, Cordya myxa, Croton obongiflorus, Curculigo orchioides, Cyperus rolundus, Digitalis parpurea, Elettaria cardomum, Tribullus terrestris, Taxus baccara, Solanum surattensis.

- **Medicinal Plants Cultivated in Nilgiri (T. Sekar Ind. For 2004 January) :**

  Plants Cultivated at Dodabela

  Resemarinus officinalis- labiatae - Skui, Uses eczema, dandruff, bronchitis rhenralsin.

Thymus vulgaris		
Pelagonium graveolens	Geraniaceae	Perfume
Oreganum vulgare	Lamiaceae	Food flavour
Petroselinum crispum	Umbellifera	Sedative
Salvia officinalis	Lamiaceae	Tonsilitis
Tanacetum parthenum	Asteraceae	Migrane

- **Cultivation of other Species all over the Country.**

  Piper longum, Cardamon, Clove, Natmeg, Cinnamom, Ginger, Termuric, Mentha arvensis, Mentha piperata, Cymbopogon, C. martini, Vetiveria zizamioides, Eucalyptus cotriodora, Santalum album, Oseimum gratissimum, Jasminum sp., Geraneum sp., Saussurea sp., Cyperus rotundus, Skimmia laurala, Artopha belladonna, A. acuminata, Cloriosa superba, Colchicum autumnate, Caltha ranthus, Rauwolfia senpentiva, Cephaelis ipececauhna, Ammi majus, Pyrethrum,

Crysanthmum cenariaefolium, Cinchona officinalis, C. muramilka, C. robusta, C. calisya, C. ledgerina, Digitalis sp., Carthmus sp., Datura metal, Solanum bhasianum, Costas speciosus, Dioscorea prazeri, D. deltoidea.

- **Eastern Himalayan sp. which have immense prospects for large scale cultivation (R.C. Sundruyal, Ind. For. Jan, 2005)**

  Aconitum ferox, A. hetrophyllum, Acorus catamus, Amonum subulatum, Androgrophis culata, Aquilaria malaccensis, Cinnamomum zeylanicum, Coptis teeta, Dioscorea flroribunda, Gymnaderia orclidis, Illicicum griffithii. Oroxylum indicum, Panax pseudo ginseng, Picororhiza kurroa, Piper lomgum, Rubia cordifolia, Taxus baccata, Swertia cirata, Tinospora cordifolia, Whitharia somnifera.

- The contribution of Dr. Wallich, Dr. Wight, Dr. Birdwood, Dr. Baden Powel, Dr. Kanailal Dey, Sir George King, Dr. Kirtikar, Major Basu, Dr. Chopra, Dr. Nadkarni, Kaviraj Braja Chandra Gupta are praise worthy.
- Dr. Chopra mentioned indiscriminate exploitation of plants like Rauwolfia, Belledona and many other medicinal plants.
- Dr. Chopra feels that cultivation of Digitalis, Belladona, Hyoseyamus, Pyrethrum, Seena, Wattle, Derris, Geranium, Pepperment, Datura and several other species has been very positive and encouraging step.
- Since 1947 India has made tremendous progress in agrotechnology, process-technology, standardization quality control, research and development.
- Leading Indian research centers have started to seek patentable genes using the latest genetic engineering techniques. More than 400 useful plants have been identified.

It is an undisputed fact that India is one of the richest in medicinal plants possessing diverse medicinal properties. Various qualities of medicinal plants for remedy of various diseases. The history on researches on medicinal plants and various information related to the subject has not been discussed in this work as such information are available in various literatures.

The most acceptable work of the authorities on the subject namely Kirtikar and Bose, Ashima Chatterjee, R.N. Chopra, K.P. Biswas, Sibakali Bhattacharya and a few others who have done indepth research on the qualities of medicinal plants their chemistry and applications are available in various literatures. Nevertheless these literatures hardly mention about the present status of medicinal plants of this country. They do not even mention about the threats the medicinal plants are facing depletion these days.

Wide Scale Cultivation is the Remedy for Rapid Depletion of Species

The supply base of ninety percent herbal raw drugs used in the manufacture of Ayurveda, Siddha, Unani and Homoeopathy systems of medicine is mainly from the forests. Besides this, plants are also used in various industries producing herbal items other than medicines. This wild source is speedily sinking day by day.

Therefore, there is need for conservation and sustainable use of medicinal plants. Cultivation is clearly a sustainable alternative to the present collection of medicinal plants from the wild. This can be a potential provider of returns to the farmers/cultivators.

Keeping the aforesaid view of R.B.S. Rawat of National Medicinal Plants Board, the author thinks there is need for a very wide scale projects for cultivation of at least 300 species of medicinal plants in India.

Chapter Fifteen
Conservation Strategy

Conservation Strategy

The Indian Forest Act, 1972 took care of increasing demand for wild medicinal plants in the organized manufacturing sector. Under this Act the export of *Rauwolfia serpentina* has been banned.

The following plants have been over exploited :

> *Clausena excavata, Atylosia scarabaeoides, Andrographis painculata, Helecteres isora, Centrella asiaticia, Jatropa curcas, Anisomeles ovata, Scoparia dulcis, Plumeria acutifolia, Vitis repanda, Nyctanthes arbortristis, Phyllanthus niruri, Echinocloa colona, Feronia elephantum, Abrus precatorius, Erycibe panieulata, Recellia suffruticosa, Vernonia anthelmintica, Acorus calamus, Orchis laxiflora, Asparagus racemosus, Polygala crota lariodis.*

Dhan Singh *et al.* (Ind. For. March 2005) writes that commercial enterprises and local dwellers are regularly exploiting natural heritage of medicinal plants in Uttaranchal. So, there is an urgent need of conservation of these valuable plants through cultivation. Poor marketing structure in the country is the primary challenge towards its promotion and cultivation.

Uttaranchal Strategy

There is no prohibition in the collection of the following species:

> *Azadirachta indica* (Neem), *Boerhaavia diffusa* (Punarnava), *Calotropis* sp. (Aak), *Cymbopogon martini* (Agyadhas), *Eclipta alba* (Mrigraj), *Fumaria vallantii* (Shahtaria), *Hibiscusrosa sinensis* (Jaba), *Mentha viridis* (Pudina), *Mimosa pudica* (Lajjavati), *Nymphaea* sp. (Padma), *Ocimum sanctum* (Tulsi), *Phaseolus trilobus*

(Moongparni), *Phyllanthus niruri* (Bhuinaonla), *Ricinus communis* (Arand), *Rosa sp.* (Golab), *Sida cordifolia* (Bala), *Abutilon indicum* (Alivala), *Solanum nigran* (Makay), *Tephrosia purpurea* (Sarpunkha), *Tribulus terrestris* (Chota gokhra). These are all growing wild but there is no inventory.

(Anil Kumar et., al. Ind. For. 2004 Jan.)

Summary of the Responsibilities of the Various Ministries and Departments of Government of India Regarding Medicinal Plants

Ministry/Department	Subject/Area of Work
Department of Indian Systems of Medicine	Preparation of list of medicinal plants in ISM. Documentation of local health traditions and Indian System of medicine and homeophthy. Encouragement to ex-situ cultivation. Development of agro technologies.
Department of Bio-technology.	Tissue culture and preservation of medicinal plants.
Department of Science and Technology	Bio-technologies, agro-technologies, CSIR germplasm preservations, etc.
Minisrty of Agriculture	Ex-situ propagation of medicinal plants. Development of agro-technologies. Tissue culture and preservation of medicinal plants.
MInistry of Environment and Forests	Conservation of medicinal plants. Identification and notification of threatened species and advice to the Ministry of Commerce to regulate their export. Documentation of ethno-botanical use of medicinal plants. Studies on ethno-biology, survey and identification of plants including medicinal plants by Botanical Survey of India.
Ministry of Commerce	Regulation of export of medicinal plants, plant products or their derivatives as per the advice of MoEF.
State Governments	Collection of medicinal plants from the wild. Ex-situ cultivation of priority species.

Summary of Responsibilities Handled by Different Wings of the Ministry of Environment and Forests Regarding Medicinal Plants

Division of MoEF	Areas of responsibility
Conservation Strategy Division	Convention of Biological Diversity (CBD). Medicinal plant conservation and protection of traditional knowledge.
	Intellectual Property Rights and community benefitsharing as required under the CBD and the proposed national legislation on biological diversity.
	Coordinating the activities of TBGRI, FRLGT, Govind Ballabh Plant Institure fo Himalayan Environment and Development and the Botanical Survey of India in the area of medicinal plants.
Forest Wing	The regulations under the Indian Forest Act 1927 and National Forest Policy 1988. ICFRE is engaged in research relating to medicinal plants.
National Afforestation and Eco Development Board (NAEB)	A scheme of NAEB on non-timber forest produce (NTFR) to promote in situ regeneration of forest produce, which includes conservation of medicinal plants to increase their production and replenish the stock.
	Local community involvement and value addition.

Towards Sustainable Management of Medicinal Plants

It is clear that in order to move towards a system of sustainable management of medicinal plants, a number of aspects need to be addressed. Some of these are summarized below.

- Better information on the current status and potential production of medicinal plants, both those that are cultivated and those that are collected from the wild, is required as a baseline from which to estimate trends in production. This is necessary before strategies for sustainable production can be developed. Current information on production potential, means for, and limits of sustainable extraction, number of units manufacturing products and their raw material requirement, use by local vaidyas, etc. needs to be brought together.
- Supply chain information is currently poor, and notoriously difficult to obtain given the non-transparent

nature of the trade. Collectors are generally not aware of the market prices of plants beyond the price paid by the local agent, and have no bargaining power. In fact, it appears that at each stage of the chain, the various actors involved have little knowledge of prices paid further along the chain. Improving the information may help the collectors to get a better share of the final price of the plants, thereby increasing their stake in sustainable management.

- There is potential for organization of collectors at the local level. Promising models for local organization of medicinal plant collectors are already being developed and may serve to reduce the risk in business and degree of dependency upon traders to which collectors are currently vulnerable. Such organization might also provide possibilities for mutually enforced codes of collection and for associated marketing benefits. Development of local institutions, with external facilitation and micro-credit assistance provided to primary collectors will support the development of micro-enterprises. This may enhance the bargaining power of primary collectors and shorten the supply chain- although it is likely that local agents will still have a role to play. The Small Industrial Development Bank of India (SIDBI) has several such schemes of extending micro - credit facilities for enterprise develpment. Small -scale value addition options, which can be carried out at primary collector's level and community level, will yield better results and ensure sustainable management and development of resources.
- The legal and administrative structure pertaining to medicinal plants can also play an important role in sustainable management. Regulatory mechanisms that control the extent and nature of extraction can ensure that plants are sustainably harvested, while government support prices (or other incentives) can help ensure a fair share for the primary collector/cultivator.
- Means for ensuring quality are of concern to the industry and to consumers. There is a need for some system of quality control to be developed. Apart from ensuring the

quality of raw material of correct botanical specifications, it is also necessary that the ingredients listed on ayurvedic products is actually used in the specified proportions.

- Another quality related aspect pertains to the preservation of the harvested raw material. Many species are harvested during the monsoon period and the moisture in the atmosphere makes the harvested raw material susceptible to fungal attacks. Currently there are practically no facilities or methods available at the collector level that prevent the raw drug from getting a fungal attack. The raw material either gets thrown away or, as is often the case, infected raw material is used in the final formulation. Any research undertaken on the preservation of the raw drug will hence go a long way in quality- and waste-control.
- The increasing involvement of casual untrained labour in the collection of medicinal plants from the wild is resulting in the use of unscientific harvest practices that are damaging to the plant as well as the environment. There is a need to not only impart scientific training to harvesters but also to educate them about the short-term and long-term advantages of following harvest practices that do not damage the plants in the long run.
- Support to small and marginal farmers to undertake cultivation of low-risk medicinal plants cannot only help bring marginal lands under cultivation but also increase production as well as improve returns to these farmers. This will, at least in some cases, reduce pressure on forest areas to meet income needs of the dependent communities.

Medicinal Plants: Needs More of Conservation and Preservation:

There should be a number of herbaria where most of the species of medicinal plants will be preserved in sheets and listed. Various herbaria viz. sibpur Botanical Garden at Howrah under B.S.I., Indian council of Forest Research and Education, Dehra Dun; Central Drug Research Institute, Lucknow; Drug Research Laboratory in Jammu and Kashmir have a good collection of medicinal plants of India. Fruits and seeds of all species are to be properly preserved in formalin and in other chemicals. This needs proper planning. The regional herbaria of B.S.I. can be a good resource base, as correct identification of species is of prime importance.

It is worthwhile in this connection to mention about vast literature and preserved materials and manuscripts written on palm leaves in the past which were mutilated by the invaders. The conquerors have introduced their own system of medicines as they were hostile towards the systems by the alien rulers. For about a thousand years Indian research, as a result, suffered a set back.

Appendix

List of some important publications.

Sl.No.

1. **A report of medicianl plants of Kachchh** (Gujarat) - CCRAS, 61–65, Institutional Area, Janakpuri, New Delhi-58, 1998.
2. **Contribution of medico-botany of east Godavari and west Godavari district of Andhra Pradesh** -CCRAS, 61–65, Institutional Area, Janakpuri, New Delhi-58, 1989.
3. **Glimps of medico-botany of Bastar district (Madhya Pradesh)** - CCRAS , 61–65, Institutional Area, Janakpuri, New Delhi-58, 1990.
4. **Medico-Botanical exploration of Puri district (Orissa)-** - CCRAS, 61–65, Institutional Area, Janakpuri, New Delhi-58, 1989.
5. **Medico-Ethno-Botany of Sonebhadra district-** CCRAS, 61–65, Institutional Area, Janakpuri, New Delhi-58, 1993.
6. **Medico-Ethno-Botanical exploration of Sikkim Himalayas-** CCRAS, 61–65, Institutional Area, Janakpuri, New Delhi-58, 1991.
7. **Medical Plants of Nagpur and Wardha forest division (Maharashtra)-** CCRAS,61–65, Institutional Area, Janakpuri, New Delhi- 58, 1999.
8. **Observation of Medico-Botany of Andaman- Nicobar Islands-** CCRAS, 61–65, Institutional Area, Janakpuri, New Delhi -58, 1980.
9. **Preliminary techno Economical Survey of natural resources and herbal wealth of Laddakh-**CCRAS, 61–65, Institutional Area, Janakpuri, New Delhi-58, 1978.
10. **Tribal pocket of Nilgiris recording of the field study on medicinal flora and health practices-**CCRAS, 61–65, Institutional Area, Janakpuri, New Delhi-58, 1976.
11. **Uttarkhand vanoushadhi Darshika-** CCRAS,61–65, Institutional Area, Janakpuri, New Delhi-58, 1977.

12. **Cultivation of Guggulu**-CCRAS, 61–65, Institutional Area, janakpuri, New Delhi-58, 1999.

13. **Experimental Cultivation of Saffron (Kumkum)**-CCRAS, 61–65, Institutional Area, Janakpuri, New Delhi-58, 1995.

14. **Pharmacognosy of Indigenous drugs**-CCRAS, 61–65, Institutional Area, Janakpuri, New Delhi-58, 1999.

15. **Phytochemical investigation of certain medical plants used in Ayurveda**-CCRAS, 61–65, Institutional Area, Janakpuri, New Delhi-58, 1990.

16. **Database on medicinal plants used in Ayurveda Volume-I,II& III**- CCRAS, 61–65, Institutional Area, Janakpuri, New Delhi-58, 2000.

17. **Album ISM medicinal plants -PLIM, 111-A.C.G.O. Complex-1,** Kamla Nehru Nagar Ghaziabad, 1997.

18. **Album of crude drugs-** PLIM, 111-A.C. G.O. Complex- 1, Kamla Nehru Nagar, Ghaziabad, 1999.

19. **Plants drugs of Ayurvedic pharmacopoeia of India Volume-1-PLIM,111-A.C.G.O. Complex-1,** Kamla Nehru Nagar Ghaziabad, 2001.

20. **A contribution of medicinal plants of Aligarh (Uttar Pradesh) -1** CCRUM, 61–65, Institutional Area, Janakpuri, New Delhi - 58.

21. **Medicinal plants of Gwalior forests division**-CCRUM, 61–65, Institutional Area, Janakpuri, New Delhi -58.

22. **Medicinal Plants of Andhra Pradesh Part-1**-CCRUM, 61–65, Institutional Area, Janakpuri, New Delhi-58.

23. **Medicinal Plants of North Arcot district, Tamil Nadu**-CCRUM, 61–65, Institutional Area, Janakpuri, New Delhi-58.

24. **Potential antimalarial herbal drugs from south eastern Indian** -CCRUM, 61–65, Institutional Area, Janakpuri, New Delhi-58.

25. **A guide to important medicinal plants used in Homoeopathy Volume-1**-HPL,111-A C.G.O. Complex-1, Kamla Nehru Nagar Ghaziabad, 1996.

26. **A guide to important medicinal plants used in Homoeopathy Volume-II**-HPL, III-A C.G.O. Complex -1 Kamla Nehru Nagar Ghaziabad, 1997.

27. **A photographic album on medicinal plants used in Homoeopathy, Volume-I-** HPL , III-A C.G.O. Complex -1, Kamla Nehru Nagar Ghaziabad. 1998.

28. **A photographic album on medicinal plants used in Homoeopathy, Volume-II**-HPL, III-A C.G.O. Complex -1, Kamla Nehru Nagar Ghaziabad, 1999.

29. **A compendium of active principles/phytochemicals of medicinal plants used in Homoeopathy, Volume-I-** HPL, III-A C.G.O. Complex-1, Kamla Nehru Nagar Ghaziabad, 2001.

30. **A check list of Homoeopathic medicinal plants of India-** CCRH,61–65, Institutional Area, Janakpuri, New Delhi- 58, 1995.

31. **Common Indian plants used in Homoeopathy** - CCRH, 61–65, Institutional Area, Janakpuri, New Delhi-58, 1998.

32. **A series of 25 medicinal plants by D.N. Tiwari, et.al.** -Uttan Centre for Sustainable Development & Poverty Alleviation 18-A, Auckland Road, Allahabad-2001.

(*Source* : National Medical Plants Board)

Bibliography

1. Anonymous (1999)– State of Forest Report, F.S.I. India.
2. Aryal, M. (1993)– The Trade in Himalayan herb.
3. Basu, B.D. (1997)– Indian Medicinal Plant.
4. Biswas, K.P. & Ghosh A. (1977) – Bharatiya Vanausadh
5. Bodding P.O.(1988) –Studies in Santhal Medicine.
6. Botanical Survey of India (1983) – Flora and vegetable of India.
7. Chopra R.N. *et. al.* (1956) – Glossary of Indian medicinal plants CSIR, New Delhi.
8. Chaudhuri R.D. (1932) – Herbal drug Industry.
9. Chaudhuri, A.B (1991) – Environment and Ecology of Herb-Shrub Flora.

   Chaudhuri, A.B. *et. al.* (2002) – Biodiversity endangered.

   Chaudhuri, A.B. *et. al.* (2003) – Megadiversity conservation.
10. Chopra, R.N. *et. al.* (1958) – Indigenous drug of India.
11. Handa, S.S. (1992) – Medicinal plants based drug.
12. Kamboj, V.P. (2000) – Herbal medicine, current science.
13. Kirtikar & Bose (1918) – Indian Medicinal Plant.
14. Mudgal, V., Hazra, P.K. (1997) – Floristic diversity and Conservation.
15. Mukherjee, G.N. (1974)– History of Indian Medicine.
16. Mulliken T.n.d. - New Support for medicinal plants.
17. National Medical Plant Board (2000) – Cultivation Practices of some commercially important medicinal plants.
18. Niraj Subrat *et. al.* (2002) – The ayurvedic medicine industry.
19. Oliver - Bever B. (1986) – Medicinal Plants of Tropical Africa.
20. Planning Commission (2000) – Report of the Task Force on Conservation and Sustainable use of medicinal plants.

21. Suma, T.S. (1998) – The medicinal plants are melting.

22. Srivastava J. *et. al.* (1996) – Medicinal plants.

23. Turtha (1988) – The ayurveda encyclopedia.

Index